Lecture Notes in Mathematics

Edited by A. Dold and B. Eckmann
Series: Forschungsinstitut für Mathematik, ETH Zürich

913

Olli Tammi

Extremum Problems for Bounded Univalent Functions II

Springer-Verlag
Berlin Heidelberg New York 1982

Author

Olli Tammi
Department of Mathematics, University of Helsinki
Hallituskatu 15, 00100 Helsinki 10, Finland

AMS Subject Classifications (1980): 30C20, 30C50, 30C55, 30C75.

ISBN 3-540-11200-6 Springer-Verlag Berlin Heidelberg New York
ISBN 0-387-11200-6 Springer-Verlag New York Heidelberg Berlin

Printed in Germany

Printing and binding: Beltz Offsetdruck, Hemsbach/Bergstr.
2141/3140-543210

Preface

"Extremum Problems for Bounded Univalent Functions", Volume 646 of Lecture Notes in Mathematics, was published in 1978. Its aim was to found such generalized Grunsky type inequalities which allow direct sharp estimation of functionals so chosen that equality is reached by certain prescribed solutions of Schiffer's differential equation. Volume 646 constitutes an introduction to the problematics mentioned. Actually, there are no tests in it which could indicate the effectiveness of the ideas proposed.

The present work is devoted to the tests and is thus a continuation to the previous one. The results are due to discussions with colleagues and students belonging to our research group in Helsinki. The present development has benefited in an essential way from the ideas of H. Haario, O. Jokinen and R. Kortram, as can be seen from the short reference list.

The estimation technique developed appears to be effective in problems connected with the first non-trivial coefficient body (a_2, a_3) of bounded univalent functions $S(b)$. The next body, (a_2, a_3, a_4), seems to be just on the limit of the range of effectiveness of our tool. In the real subclass $S_R(b)$ of $S(b)$ one finds a complete characterization of the algebraic part of the coefficient body. As a consequence of this the maximum of a_4 can be determined in $S_R(b)$ for all values of b.

The present computer technique allows illustrating some of the results by graphs unattainable by purely manual computations. I am grateful for these graphs to O. Jokinen who skillfully composed the drawing programs involved.

Helsinki, March 1981

Olli Tammi

Contents

V THE FIRST COEFFICIENT BODY AND RELATED PROBLEMS IN $S(b)$

1 § Determination of the Coefficient Body (a_2, a_3) by Aid of a Perfect Square Method

1. Introduction

In Paragraph IV.2 of [1] Schiffer's differential equation was used for predicting the existence of inequalities which would generalize the Power - inequality. This is a Grunsky type inequality for the class $S(b)$ of bounded univalent functions. - In all terms and notations used here without any closer specification we refer to those in [1].

In [2] H. Haario and O. Jokinen show, how certain expected extensions can be found by aid of a unified and elementary method, by estimating proper perfect square presentations connected to a functional obtainable from Schiffer's differential equation. It appears that this perfect square method is able to give inequalities which completely characterize the first non-trivial coefficient body, to be denoted here by (a_2, a_3). Moreover, it turns out that certain functionals pointed out by G. Schober in [3] can be handled by using the same method and related inequalities. Hence, we have a good reason to treat these problems as an entity in the three paragraphs of this chapter V, forming a natural continuation to the four chapters of [1].

In Chapter VI Löwner-identities are further used in the real class $S_R(b) \subset S(b)$ in rederiving and generalizing the Power-inequality so that an extended information of the second coefficient body can be obtained.

2. The Boundary Functions 2:2

Let us consider, as in [1], bounded and normalized univalent functions $f \in S(b)$ for which we assume:

$$\begin{cases} z \in U = \{z \in C \mid |z| < 1\}, \\ f(z) = b(z + a_2 z^2 + \dots), \\ |f(z)| < 1, \quad b \in (0,1]. \end{cases}$$

We start from the Löwner-formulae (cf. [1], (13)/I.2.8, p. 33)

$$a_2 = -2\int_b^1 \kappa du, \quad \delta = a_3 - a_2^2 = -2\int_b^1 u\kappa^2 du, \quad \kappa = e^{i\vartheta},$$

which give the following identity for any $x_0 \in C$:

$$\delta + x_0 a_2 = -2\int_b^1 (u\kappa^2 + x_0\kappa)du$$

$$= -2\int_b^1 u(\kappa^2 + \frac{x_0}{u}\kappa + \frac{x_0^2}{4u^2} - \frac{x_0^2}{4u^2})du$$

$$= -2\int_b^1 u(\kappa + \frac{x_0}{2u})^2 du - \frac{x_0^2}{2}\log b;$$

$$\begin{cases} \delta + x_0 a_2 + \dfrac{x_0^2}{2}\log b = -2\displaystyle\int_b^1 A^2 du, \\ A = \sqrt{u}(\kappa + \dfrac{x_0}{2u}) = X + iY. \end{cases}$$

From this initial identity we deduce:

$$\text{Re}\,(\delta + x_0 a_2 + \frac{x_0^2}{2}\log b) = -2\int_b^1 \text{Re}\,A^2 du$$

$$= 2\int_b^1 |A|^2 du - 4\int_b^1 X^2 du$$

$$= 1 - b^2 - \frac{|x_0|^2}{2} \log b - \operatorname{Re}(\bar{x}_0 a_2) - 4 \int_b^1 X^2 du.$$

Our first estimation shall be based on this identity:

$$(1) \quad \begin{cases} \operatorname{Re}\left(\delta + x_0 a_2 + \frac{x_0^2}{2} \log b\right) = 1 - b^2 - \frac{|x_0|^2}{2} \log b - \operatorname{Re}(\bar{x}_0 a_2) - 4 \int_b^1 X^2 du, \\ X = \sqrt{u}\left(\cos \vartheta + \frac{\operatorname{Re} x_0}{2u}\right). \end{cases}$$

In order to estimate $\operatorname{Re} \delta$ upwards we deduce for it from (1):

$$(2) \qquad \operatorname{Re} \delta = -\operatorname{Re}(x_0 a_2) - \frac{1}{2} \log b \operatorname{Re}(x_0^2) + 1 - b^2 - \frac{|x_0^2|}{2} \log b - \operatorname{Re}(\bar{x}_0 a_2)$$

$$- 4 \int_b^1 X^2 du = F - 4 \int_b^1 X^2 du,$$

$$F = 1 - b^2 - \operatorname{Re}\{(x_0 + \bar{x}_0) a_2\} - \frac{1}{2} \log b \operatorname{Re}\{x_0^2 + |x_0|^2\}$$

$$= 1 - b^2 - 2 \operatorname{Re} x_0 \cdot \operatorname{Re} a_2 - \log b \, (\operatorname{Re} x_0)^2$$

$$= 1 - b^2 + \frac{(\operatorname{Re} a_2)^2}{\log b} - \log b \left(\operatorname{Re} x_0 + \frac{\operatorname{Re} a_2}{\log b}\right)^2.$$

Clearly, we obtain the best possible inequality for $\operatorname{Re} \delta$ by choosing $\operatorname{Re} x_0$ so that F is minimized. This <u>optimal choice</u> of x_0 is thus

$$\operatorname{Re} x_0 = - \frac{\operatorname{Re} a_2}{\log b};$$

$$F = 1 - b^2 + \frac{(\operatorname{Re} a_2)^2}{\log b}.$$

This value of F gives an upper bound for $\operatorname{Re} \delta$, as is seen from (2):

$$\operatorname{Re} \delta \leq F = 1 - b^2 + \frac{(\operatorname{Re} a_2)^2}{\log b}.$$

Equality here is reached provided that it can be reached in the estimation

$$-4\int_b^1 X^2 du \leqq 0,$$

i.e. if we can find a κ for which $X \equiv 0$. For such κ there holds necessarily

$$\cos\vartheta = -\frac{\operatorname{Re} x_0}{2u} = \frac{\operatorname{Re} a_2}{2u \log b}. \tag{3}$$

This implies, because $\cos\vartheta(u) \in [-1,1]$, restrictions for $\operatorname{Re} a_2$ as follows.

$\underline{\operatorname{Re} a_2 \geqq 0}$

$$-1 \leqq \frac{\operatorname{Re} a_2}{2b \log b}$$

⇒

$$-2b \log b \geqq \operatorname{Re} a_2 \geqq 0$$

⇒

$$|\operatorname{Re} a_2| = \operatorname{Re} a_2 \leqq 2b\ |\log b|.$$

$\underline{\operatorname{Re} a_2 \leqq 0}$

$$\frac{\operatorname{Re} a_2}{2b \log b} \leqq 1$$

⇒

$$\operatorname{Re} a_2 \geqq 2b \log b$$

⇒

$$|\operatorname{Re} a_2| = -\operatorname{Re} a_2 \leqq 2b\ |\log b|.$$

Observe that in $S(b)$

$$\operatorname{Re} a_2 \in [-2(1-b), 2(1-b)]$$

and for any such a_2 we can, of course, choose

$$\operatorname{Re} x_0 = -\frac{\operatorname{Re} a_2}{\log b},$$

which implies that (2) gives for an arbitrary a_2 the true inequality

(4) $$\operatorname{Re} \delta \leqq 1 - b^2 + \frac{(\operatorname{Re} a_2)^2}{\log b}.$$

However, reaching the equality here requires reaching it in

$$-4 \int_b^1 u\left(\cos \vartheta - \frac{\operatorname{Re} a_2}{2u \log b}\right)^2 du \leqq 0,$$

which implies that (4) can be sharp only for those numbers a_2 for which

(5) $$|\operatorname{Re} a_2| \leqq 2b\,|\log b|.$$

On the other hand, for those numbers a_2 the equality function $\cos \vartheta$ is, according to (3), defined on the whole interval $b \leqq u \leqq 1$. - Let us consider more closely that $\kappa = e^{i\vartheta}$ which is defined by the equality condition (3).

$\operatorname{Re} a_2 \leqq 0.$

For brevity, let us denote

$$\begin{cases} \sigma = -\dfrac{\operatorname{Re} x_0}{2} = \dfrac{\operatorname{Re} a_2}{2 \log b} \in [0,b]; \\ \operatorname{Re} x_0 \cdot \operatorname{Re} a_2 \geqq 0; \end{cases}$$

(6) $$\cos \vartheta = \frac{\sigma}{u}, \quad b \leqq u \leqq 1.$$

Further, let

$$a_2 = U + iV.$$

For U we have in the extremal case:

$$U = -2 \int_b^1 \cos \vartheta \, du = -2 \int_b^1 \frac{\sigma}{u} \, du = 2\sigma \log b.$$

Because we can take in the extremal case

$$(7)\qquad \sin\vartheta = \begin{cases} \sqrt{1-\frac{\sigma^2}{u^2}}, & b \leqq u \leqq c, \\ -\sqrt{1-\frac{\sigma^2}{u^2}}, & c \leqq u \leqq 1, \end{cases}$$

we obtain for V belonging to an extremal function:

$$V = -2\int_b^1 \sin\vartheta\, du = -2\int_b^c \sqrt{1-\frac{\sigma^2}{u^2}}\, du + 2\int_c^1 \sqrt{1-\frac{\sigma^2}{u^2}}\, du.$$

Hence, for this V there holds:

$$(8)\qquad \left|\frac{V}{2}\right| \leqq \int_b^1 \sqrt{1-\frac{\sigma^2}{u^2}}\, du = \sqrt{1-\sigma^2} - \sigma\, \overline{\text{arc}}\cos\sigma - \sqrt{b^2-\sigma^2} + \sigma\, \overline{\text{arc}}\cos\frac{\sigma}{b},$$

where the equality belongs to the limit case $c = b$ of the extremal mapping of the type 2:2, defined by (6) and (7). – We refer here to III.3.2-III.3.5 of [1], where the corresponding boundary functions are studied in detail.

Observe, that we could let $\sin\vartheta$ change its sign in more than one point on $[b,1]$. Because

$$\text{Re}\,\delta = 1 - b^2 - 4\int_b^1 u\cos^2\vartheta\, du$$

is independent of the choice of these points, we prefer (7) as the simplest possibility for the extremum function. – As a matter of fact, the extremum function appears to be independent of the points where $\sin\vartheta$ changes its singn. The considerations needed to check this are omitted. This is justified especially because all estimations based on Löwner's method generally leave open the uniqueness of the extremal functions in $S(b)$, although the inequalities found are sharp there.

<u>$\text{Re}\, a_2 \geqq 0$.</u>

Now denote

$$\begin{cases} \sigma = \dfrac{\text{Re}\, x_0}{2} = -\dfrac{\text{Re}\, a_2}{2\log b} \in [0,b]; \\ \text{Re}\, x_0 \cdot \text{Re}\, a_2 \geqq 0; \end{cases}$$

$$\cos\vartheta = -\frac{\sigma}{u}.$$

Again, $\sin\vartheta$ is defined by (7). In the present case

$$U = -2\sigma \log b$$

and for V the condition (8) remains to hold.

Theorem 1. Denote $\delta = a_3 - a_2^2$, $a_2 = U + iV$. Then in $S(b)$

(9) $$\operatorname{Re}\delta \leqq 1 - b^2 + \frac{U^2}{\log b}.$$

This condition is sharp for

(10) $$|U| = |\operatorname{Re} a_2| \leqq 2b\,|\log b|$$

for an extremal $\vartheta = \vartheta(u)$, $b \leqq u \leqq 1$, defined by

(11) $$\frac{U}{|U|}\cos\vartheta = -\frac{\sigma}{u},\quad \sin\vartheta = \begin{cases} \sqrt{1-\frac{\sigma^2}{u^2}}, & b \leqq u \leqq c,\\ -\sqrt{1-\frac{\sigma^2}{u^2}}, & c \leqq u \leqq 1,\end{cases}$$

(for $U = 0$ take $\frac{U}{|U|} = \pm 1$) where

(12) $$\sigma = \left|\frac{U}{2\log b}\right| \in [0,b].$$

The coefficient $a_2 = U + iV$ of the extremal function f is limited so that

(13) $$\begin{cases} U = \mp 2\sigma \log b,\\ |V| \leqq 2\left(\sqrt{1-\sigma^2} - \sigma\,\overline{\operatorname{arc}}\cos\sigma - \sqrt{b^2-\sigma^2} + \sigma\,\overline{\operatorname{arc}}\cos\frac{\sigma}{b}\right).\end{cases}$$

σ and the parameter x_0 of the initial identity (1) are connected by the conditions:

(14) $$\begin{cases} U \leqq 0:\ \sigma = -\dfrac{\operatorname{Re} x_0}{2} = \dfrac{U}{2\log b},\\ U \geqq 0:\ \sigma = \dfrac{\operatorname{Re} x_0}{2} = -\dfrac{U}{2\log b}.\end{cases}$$

It is useful to point out, that in utilizing (2) we may avoid the above optimization of x_0 by proceeding as follows. (2) gives for an arbitrary x_0:

$$\mathrm{Re}\ \delta \leqq F = 1 - b^2 - 2\ \mathrm{Re}\ x_0 \cdot \mathrm{Re}\ a_2 - \log b \cdot (\mathrm{Re}\ x_0)^2.$$

Equality here can be reached if and only if $X \equiv 0$, i.e.

$$\cos \vartheta = -\frac{\mathrm{Re}\ x_0}{2u}.$$

Thus, a necessary condition for x_0 defining the existence of such ϑ is

$$\mathrm{Re}\ a_2 = -2 \int_b^1 \cos \vartheta \ du = \int_b^1 \frac{\mathrm{Re}\ x_0}{u}\ du = -\log b \cdot \mathrm{Re}\ x_0$$

$\Rightarrow$

$$\mathrm{Re}\ x_0 = -\frac{\mathrm{Re}\ a_2}{\log b}.$$

This is the same choice which was found earlier by maximizing the latter part $-4 \int_b^1 X^2 du$ of the sum (2). - In the optimization procedure the former part F of (2) was minimized. The fact that the maximum and minimum mentioned occur simultaneously is necessary for the successful utilization of the identity (2). It appears that also for the identities which are applicable for the remaining cases a similar situation holds.

Let us consider the equality case more closely. According to the initial identity

$$\delta + x_0 a_2 + \frac{x_0^2}{2} \log b = -2 \int_b^1 A^2 du$$

we have in the equality case

$$\mathrm{Im}\ (\delta + x_0 a_2 + \frac{x_0^2}{2} \log b) = -4 \int_b^1 XY du = 0.$$

Thus, in the equality case of the condition obtained from (1)

$$\operatorname{Re}\,(\delta + x_0 a_2 + \frac{x_0^2}{2}\log b) \leqq 1 - b^2 - \frac{|x_0|^2}{2}\log b - \operatorname{Re}\,(\bar{x}_0 a_2)$$

there holds

$$\begin{cases} \delta + x_0 a_2 + \dfrac{x_0^2}{2}\log b = 1 - b^2 - \dfrac{|x_0^2|}{2}\log b - \operatorname{Re}\,(\bar{x}_0 a_2), \\ \operatorname{Re}\, x_0 = -\dfrac{\operatorname{Re}\, a_2}{\log b}. \end{cases}$$

By using the notations

$$a_2 = U + iV, \quad x_0 = -C = -(C_1 + iC_2),$$

we have

$$C_1 = \frac{U}{\log b};$$

$$\delta - C_1U + C_2V - i(C_2U + C_1V) + \frac{C_1^2 - C_2^2 + 2iC_1C_2}{2}\log b$$

$$= 1 - b^2 - \frac{C_1^2 + C_2^2}{2}\log b + C_1U + C_2V;$$

$$\delta = 1 - b^2 + 2C_1U - C_1^2\log b + i[(U - C_1\log b)C_2 + C_1V];$$

$$(15) \qquad \delta = 1 - b^2 + \frac{U^2}{\log b} + i\,\frac{UV}{\log b} = \delta^\circ + R$$

where

$$\begin{cases} \delta^\circ = \dfrac{a_2^2}{2\log b}, \\ R = 1 - b^2 + \dfrac{|a_2|^2}{2\log b}. \end{cases}$$

This shows that $\delta - \delta^\circ = R$ is real and positive in the equality case of (9), as was to be expected.

In order to generalize the inequality (9) we may apply the rotation where, by aid of a unit vector τ $(|\tau| = 1)$, $f(z)$ is replaced by

$$\begin{cases} \tau^{-1}f(\tau z) \in S(b), \\ (a_2,\delta) \text{ of } f(z) \Rightarrow (\tau a_2, \tau^2\delta) \text{ of } \tau^{-1}f(\tau z). \end{cases}$$

(9) assumes the form

$$\text{Re }(\tau^2\delta) \leqq 1 - b^2 + \frac{(\tau a_2 + \tau^{-1}\bar{a}_2)^2}{4\log b}$$

and (15) gives in the equality case of this:

$$\tau^2\delta = \tau^2 \frac{a_2^2}{2\log b} + R$$

i.e.

(16) $$\delta = \delta^0 + R\tau^{-2}.$$

Thus we see that δ for a fixed τ is restricted in a half-plane. When τ is used as a free parameter we obtain as an intersection of these half-planes a disc, the boundary of which is determined by (16). This can be verified by starting from (9), which gives

$$\text{Re }(\delta - \delta^0) \leqq 1 - b^2 + \frac{(\text{Re } a_2)^2}{\log b} - \frac{\text{Re }(a_2^2)}{2\log b}$$

$$= 1 - b^2 + \frac{2U^2 - U^2 + V^2}{2\log b} = 1 - b^2 + \frac{|a_2|^2}{2\log b} = R.$$

The rotation $\tau^{-1}f(\tau z)$ gives

$$\text{Re }\{\tau^2(\delta - \delta^0)\} \leqq R$$

and because τ is free, this implies

$$|\delta - \delta^0| \leqq R.$$

Thus, we end up with the disc result derived in III.3.3 of [1].

In order to find out how the coefficient $a_2 = U + iV$ is restricted when

$\tau \neq 1$ is used we may introduce the notations

$$\tau = e^{iv}, \quad \tilde{a}_2 = \tau a_2 = \tilde{U} + i\tilde{V}.$$

The conditions (13) hold now for $(\tilde{U},\tilde{V})$ giving a restriction for (U,V). Let us collect the results.

Theorem 2. In $S(b)$ there holds the inequality

$$(17) \quad \mathrm{Re}\,(\tau^2\delta) \leqq 1 - b^2 + \frac{(\tau a_2 + \tau^{-1}\bar{a}_2)^2}{4\log b}$$

for any $\tau = e^{iv}$. This implies that

$$(18) \quad \begin{cases} |\delta - \delta^o| \leqq R; \\ \delta^o = \dfrac{a_2^2}{2\log b}, \\ R = 1 - b^2 + \dfrac{|a_2|^2}{2\log b}. \end{cases}$$

Equality in (17) and (18) occur for an extremal function f which has a coefficient $a_2 = U + iV$ limited by the conditions

$$(19) \quad \begin{cases} \tilde{U} = \mp 2\sigma \log b, \\ |\tilde{V}| \leqq 2(\sqrt{1-\sigma^2} - \sigma\,\overline{\mathrm{arc}}\cos\sigma - \sqrt{b^2-\sigma^2} + \sigma\,\overline{\mathrm{arc}}\cos\frac{\sigma}{b}); \\ \tilde{U} = \cos v\cdot U - \sin v\cdot V, \quad \tilde{V} = \sin v\cdot U + \cos v\cdot V, \\ \sigma \in [0,b]. \end{cases}$$

As an example we may consider the case normalized so that

$$U = a_2 = |a_2| \geqq 0.$$

This normalization appears to be general enough when illustrating different types of boundary functions. In the present case we have

$$\begin{cases} \tilde{U} = \cos v\cdot U = \mp 2\sigma \log b, \\ \tilde{V} = \sin v\cdot U; \end{cases}$$

$$(20)\quad \begin{cases} \sigma = \left|\dfrac{U\cos v}{2\log b}\right|, \\[2ex] |\sin v\cdot U| \leqq 2\left(\sqrt{1-\sigma^2} - \sigma\,\overline{\text{arc}}\cos\sigma - \sqrt{b^2-\sigma^2} + \sigma\,\overline{\text{arc}}\cos\dfrac{\sigma}{b}\right). \end{cases}$$

In [1], p. 253, the existence of the boundary function of the type 2:2 is expressed in the form

$$(21)\quad \begin{cases} t = \dfrac{\sigma}{b} \in [0,1], \\[2ex] D(\sigma) = \dfrac{\sigma}{b}\,\overline{\text{arc}}\cos\dfrac{\sigma}{b} - \sqrt{1-\dfrac{\sigma^2}{b^2}} + \sqrt{1-\sigma^2} - \sigma\,\overline{\text{arc}}\cos\sigma \\[2ex] \qquad - \dfrac{|a_2|}{2}\sqrt{1-\left(\dfrac{2\log b\cdot\sigma}{|a_2|}\right)^2} \geqq 0. \end{cases}$$

From (20) we see that

$$t = \frac{\sigma}{b} = \left|\frac{U\cos v}{2b\log b}\right| \in [0,1];$$

$$\cos v = \left|\frac{2\log b\cdot\sigma}{U}\right|;$$

$$\frac{1}{2}|U|\sqrt{1-\left(\frac{2\log b\cdot\sigma}{U}\right)^2} \leqq \sqrt{1-\sigma^2} - \sigma\,\overline{\text{arc}}\cos\sigma - \sqrt{b^2-\sigma^2} + \sigma\,\overline{\text{arc}}\cos\frac{\sigma}{b},$$

which is the condition (21). The existence conditions (19) thus agree with those found formerly by aid of the Power inequality.

We could also rederive the condition (1)/III.3.5 of [1] for boundary functions of the type 2:2 by integrating Löwner's equation for κ determined by (11). This would give an integrated condition for $\tilde{f}$ which is connected with f so that $\tilde{f}(z) = \tau^{-1}f(\tau z)$. – Later on we shall return to equations which determine extremal functions f and corresponding image domains.

3. The Boundary Functions 1:2

In the previous case 2:2 we used the choice

$$\sigma = \frac{|\mathrm{Re}\ x_0|}{2} \in [0,b]$$

in the identity (2):

$$\begin{cases} \mathrm{Re}\ \delta = F - 4 \displaystyle\int_b^1 X^2 du, \\ X = \sqrt{u}\ (\cos \vartheta \ + \dfrac{\mathrm{Re}\ x_0}{2u}) = \sqrt{u}\ (\cos \vartheta \ \mp \dfrac{\sigma}{u}). \end{cases}$$

Let us choose $\mathrm{Re}\ x_0$ so that

$$\sigma = \frac{|\mathrm{Re}\ x_0|}{2} \in [b,1].$$

Again, consider both signs of $\mathrm{Re}\ x_0$.

1) $\mathrm{Re}\ x_0 < 0$; $\sigma = -\dfrac{\mathrm{Re}\ x_0}{2}$; $X = \sqrt{u}\ (\cos \vartheta - \dfrac{\sigma}{u})$.

Now we write

$$-4 \int_b^1 X^2 du = \underbrace{-4 \int_b^\sigma u(\cos \vartheta - \frac{\sigma}{u})^2 du}_{\mathrm{I}} \underbrace{- 4 \int_\sigma^1 u(\cos \vartheta - \frac{\sigma}{u})^2 du}_{\mathrm{II}}.$$

The numbers I and II can be estimated upwards sharply.

II: $\sigma \leqq u \leqq 1 \Rightarrow \dfrac{\sigma}{u} \leqq 1.$

II $\leqq 0$, where equality is reached by choosing

$$\cos \vartheta = \frac{\sigma}{u}, \quad \sigma \leqq u \leqq 1.$$

I: $b \leqq u \leqq \sigma \Rightarrow \dfrac{\sigma}{u} \geqq 1.$

The previous choice for $\cos \vartheta$ is possible only at the point $u =$
At the other points of $b \leqq u \leqq \sigma$ $\cos \vartheta - \dfrac{\sigma}{u} < 0$. We now have

$$|\cos \vartheta - \frac{\sigma}{u}| \geqq \frac{\sigma}{u} - |\cos \vartheta| \geqq \frac{\sigma}{u} - 1 \geqq 0$$

$$-(\cos\vartheta - \frac{\sigma}{u})^2 \leqq -(\frac{\sigma}{u} - 1)^2;$$

$$I \leqq -4\int_b^\sigma u(1 - \frac{\sigma}{u})^2 du = 6\sigma^2 - 4\sigma^2 \log\sigma + 2b^2 - 8b\sigma + 4\sigma^2 \log b,$$

where equality belongs to the choice

$$\cos\vartheta \equiv 1, \quad b \leqq u \leqq \sigma.$$

Thus we have

$$-4\int_b^1 X^2 du \leqq 6\sigma^2 - 4\sigma^2 \log\sigma + 2b^2 - 8b\sigma + 4\sigma^2 \log b$$

and (2) gives hence

$$\mathrm{Re}\,\delta \leqq 1 + b^2 + 4U\sigma - 8b\sigma + 6\sigma^2 - 4\sigma^2 \log\sigma = M(\sigma).$$

Equality here is reached for

$$(22) \qquad \cos\vartheta = \begin{cases} 1, & b \leqq u \leqq \sigma, \\ \frac{\sigma}{u}, & \sigma \leqq u \leqq 1. \end{cases}$$

We can now proceed as before. σ can be optimized by requiring that $M(\sigma)$ is minimized or we determine σ so that a_2 is in accordance with (22). Both these methods lead to the same condition for σ:

$$\frac{1}{8}\frac{dM}{d\sigma} = \frac{U}{2} + \sigma - b - \sigma\log\sigma = 0;$$

$$U = -2\int_b^\sigma du - 2\int_\sigma^1 \frac{\sigma}{u}\, du = 2(\sigma\log\sigma - \sigma + b).$$

The condition found determines $\sigma = \sigma(U) \in [b,1]$ provided

$$-2(1 - b) \leqq U \leqq -2b\,|\log b| < 0.$$

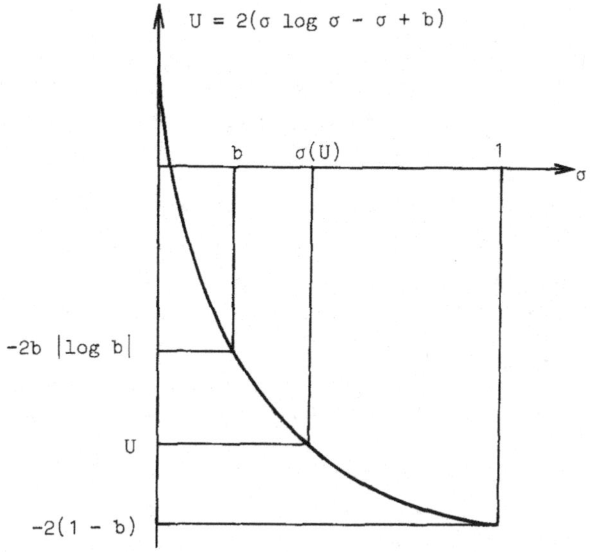

Figure 1.

For this σ we obtain

(23) $$\mathrm{Re}\ \delta \leqq \min M = M(\sigma(U)) = 1 - b^2 + 2U\sigma + 2(\sigma - b)^2 .$$

2) $$\mathrm{Re}\ x_0 > 0; \quad \sigma = \frac{\mathrm{Re}\ x_0}{2}; \quad X = \sqrt{u}\,(\cos\vartheta + \frac{\sigma}{u}).$$

In this case $-4\int_b^1 X^2 du$ is maximized for

$$\cos\vartheta = \begin{cases} -1, & b \leqq u \leqq \sigma, \\ -\frac{\sigma}{u}, & \sigma \leqq u \leqq 1; \end{cases}$$

$$-4\int_b^1 X^2 du \leqq -4\int_b^{\sigma} u(-1+\frac{\sigma}{u})^2 du.$$

Thus, this part of the upper bound of Re δ is left unchanged and in the part F the factor of U changes the sign, giving

$$\text{Re } \delta \leqq 1 + b^2 - 4U\sigma - 8b\sigma + 6\sigma^2 - 4\sigma^2 \log \sigma = M(\sigma).$$

By optimizing this with respect to σ or by requiring that

$$U = -2\int_b^{\sigma} (-1)du - 2\int_{\sigma}^1 \frac{-\sigma}{u} du$$

we obtain for σ:

$$U = -2(\sigma \log \sigma - \sigma + b).$$

Because only the sign of U is changed we find for U the limits

$$0 < 2b\,|\log b| \leqq U \leqq 2(1-b).$$

The expression (23) gives sharp upper bound for Re δ when changing the sign of U:

$$\text{Re } \delta \leqq 1 - b^2 - 2U\sigma + 2(\sigma - b)^2.$$

In both the cases 1) and 2) we have for the equality function ϑ the conditions

$$\begin{cases} \sin\vartheta = 0, & b \leqq u \leqq \sigma, \\ \sin\vartheta = \begin{cases} \sqrt{1-\frac{\sigma^2}{u^2}}, & \sigma \leqq u \leqq c, \\ -\sqrt{1-\frac{\sigma^2}{u^2}}, & c \leqq u \leqq 1. \end{cases} \end{cases}$$

Hence, there holds for V belonging to the extremal function

$$V = -2\int_b^1 \sin\vartheta\, du = -2\int_b^c \sqrt{1-\frac{\sigma^2}{u^2}}\, du - 2\int_c^1 -\sqrt{1-\frac{\sigma^2}{u^2}}\, du.$$

This gives for V the inequality

$$\left|\frac{V}{2}\right| \leq \int_{\sigma}^{1} \sqrt{1 - \frac{\sigma^2}{u^2}}\, du = \sqrt{1 - \sigma^2} - \sigma\, \overline{\text{arc}} \cos \sigma,$$

where the equality belongs to the limit case $c = \sigma$ of the extremal 1:2-mapping.

<u>Theorem 3</u>. In $S(b)$ the estimation (9) can be sharpened for the values

$$(24) \quad |\text{Re}\, a_2| = |U| \geqq 2b\, |\log b|$$

for which

$$(25) \quad \text{Re}\, \delta \leqq 1 - b^2 - 2\, |U| \sigma + 2(\sigma - b)^2.$$

Here $\sigma = \sigma(U) \in [b,1]$ is the root of the equation

$$(26) \quad \sigma \log \sigma - \sigma + b + \frac{|U|}{2} = 0.$$

Equality in (25) is reached by ϑ determined so that (for $U = 0$ take again $\frac{U}{|U|} = \pm 1$)

$$(27) \begin{cases} \frac{U}{|U|} \cos \vartheta = \begin{cases} -1, & b \leqq u \leqq \sigma, \\ -\frac{\sigma}{u}, & \sigma \leqq u \leqq 1; \end{cases} \\ \sin \vartheta = \begin{cases} 0, & b \leqq u \leqq \sigma \\ \sqrt{1 - \frac{\sigma^2}{u^2}}, & \sigma \leqq u \leqq c, \\ -\sqrt{1 - \frac{\sigma^2}{u^2}}, & c \leqq u \leqq 1. \end{cases} \end{cases}$$

The equality function f has for $a_2 = U + iV$:

$$(28) \begin{cases} U = \mp 2(\sigma \log \sigma - \sigma + b) \quad (- \text{ for } U > 0, \ + \text{ for } U < 0), \\ |V| \leq 2(\sqrt{1 - \sigma^2} - \sigma\, \overline{\text{arc}} \cos \sigma), \\ \sigma \in [b,1]. \end{cases}$$

σ is connected with x_0 so that $\sigma = \frac{|\mathrm{Re}\, x_0|}{2}$. The sign of U belonging to the extremal function is that of $\mathrm{Re}\, x_0$:

$$U \cdot \mathrm{Re}\, x_0 > 0.$$

Consider the equality case of (25) by using the initial identity which gives

$$\begin{cases} \mathrm{Im}\,(\delta + x_0 a_2 + \frac{x_0^2}{2} \log b) = -4 \int_b^1 XY du; \\ X = \sqrt{u}\,(\cos \vartheta + \frac{\mathrm{Re}\, x_0}{2u}),\ Y = \sqrt{u}\,(\sin \vartheta + \frac{\mathrm{Im}\, x_0}{2u}). \end{cases}$$

We see from (27) that $X = 0$ for $\sigma \leqq u \leqq 1$. For $b \leqq u \leqq \sigma$ we get $Y = 0$ if we choose $\mathrm{Im}\, x_0 = 0$. This is justified also because the identity (2) which gave (25) was independent of $\mathrm{Im}\, x_0$. Therefore, assume that

$$x_0 \in R$$

and obtain in the equality case of (25):

$$\begin{cases} \mathrm{Re}\, \delta = 1 - b^2 - 2\,|U|\sigma + 2(\sigma - b)^2, \\ \mathrm{Im}\, \delta = -x_0 V. \end{cases}$$

According to Theorem 3 we have

$$[b,1] \ni \sigma = \frac{|x_0|}{2} = \begin{cases} \frac{x_0}{2}, & U > 0, \\ -\frac{x_0}{2}, & U < 0. \end{cases}$$

Thus, in the equality case

$$\begin{cases} \mathrm{Re}\, \delta = 1 - b^2 - U x_0 + 2(\sigma - b)^2, \\ \mathrm{Im}\, \delta = -V x_0; \end{cases}$$

$$\delta = 1 - b^2 + 2(\sigma - b)^2 - x_0(U + iV);$$

(29) $$\delta = 1 - b^2 + 2(\sigma - b)^2 \mp 2\sigma a_2$$

(- for $U > 0$, + for $U < 0$).

We may, again, generalize the result by aid of the rotation $\tau^{-1}f(\tau z)$, $\tau = e^{iv}$, which gives $\tilde{f}$ with $\tilde{a}_2 = \tau a_2$, $\tilde{\delta} = \tau^2\delta$ and

$$\begin{cases} \tilde{U} = \dfrac{\tau a_2 + \tau^{-1}\bar{a}_2}{2} = \cos v \cdot U - \sin v \cdot V, \\[2ex] \tilde{V} = \dfrac{\tau a_2 - \tau^{-1}\bar{a}_2}{2i} = \sin v \cdot U + \cos v \cdot V, \end{cases}$$

for which Theorem 3 and (29) hold. Thus, we arrive at:

Theorem 4. In $S(b)$

$$(30)\quad \mathrm{Re}\,(\tau^2\delta) \leqq 1 - b^2 + 2(\sigma - b)^2 - 2\,|\tilde{U}|\sigma$$

$$= 1 - b^2 + 2(\sigma - b)^2 - |\tau a_2 + \tau^{-1}\bar{a}_2|\sigma$$

for any $\tau = e^{iv}$ and σ determined by

$$(31)\quad \sigma \log \sigma - \sigma + b + \frac{|\tilde{U}|}{2} = 0.$$

The values of a_2 and τ for which (30) is sharp, satisfy

$$(32)\quad |\tilde{U}| = |\cos v \cdot U - \sin v \cdot V| = \frac{|\tau a_2 + \tau^{-1}\bar{a}_2|}{2} \geqq 2b\,|\log b|,$$

$$(33)\quad |\tilde{V}| = |\sin v \cdot U + \cos v \cdot V| = \frac{|\tau a_2 - \tau^{-1}\bar{a}_2|}{2} \leqq 2(\sqrt{1 - \sigma^2} - \sigma\,\overline{\mathrm{arc}}\cos \sigma).$$

The number δ lies in the intersection of the half-planes determined by (30). The boundary curve of this intersection is given by

$$(34)\quad \delta = (1 - b^2 + 2(\sigma - b)^2)\tau^{-2} \mp 2\sigma\tau^{-1}a_2$$

(- for $\tilde{U} = \mathrm{Re}\,(\tau a_2) > 0$, + for $\tilde{U} = \mathrm{Re}\,(\tau a_2) < 0$).

As an example consider again the case

$$U = a_2 = |a_2| \geqq 0$$

and normalize, without any loss of generality,

$$\tilde{U} = |a_2| \cos v \leqq 0.$$

Thus we have, according to Theorem 4

$$\begin{cases} |\cos v \cdot |a_2|| \geqq 2b\ |lob\ b|, \\ \sigma \log \sigma - \sigma + b - \frac{\tilde{U}}{2} = 0, \\ \delta = (1 - b^2 + 2(\sigma - b)^2)\tau^{-2} + 2\sigma\tau^{-1}|a_2| = \delta_1 + i\delta_2; \end{cases}$$

$\Rightarrow$

$$(35)\quad \begin{cases} \cos v \leqq \frac{2b \log b}{|a_2|} \\ \sigma \log \sigma - \sigma + b + \frac{1}{2}|a_2||\cos v| = 0, \\ \delta_1 = 2\,|a_2| \cos v \cdot \sigma + (1 - b^2 + 2(\sigma - b)^2) \cos 2v, \\ \delta_2 = -[2\,|a_2| \sin v \cdot \sigma + (1 - b^2 + 2(\sigma - b)^2) \sin 2v], \\ E(\sigma) = \sqrt{1 - \sigma^2} - \sigma\ \overline{\text{arc}} \cos \sigma - \frac{1}{2}|a_2||\sin v| \geqq 0. \end{cases}$$

Here the control function $E(\sigma)$ is connected to $D(\sigma)$ in (21) so that

$$D(b) = E(b).$$

As an example take $b = 0.1$ and determine the domain of $a_2 = U + iV$ belonging to the extremal function for which the inequalities (9) and (25) are sharp. In Table 1 there are values giving a part of the boundary of this domain in the case

$$U \leqq 0, \quad V \leqq 0.$$

In Figure 2 the domain in question is illustrated.

Next, consider two examples

$$b = 0.1, \quad a_2 = 1.5 \quad \text{and} \quad b = 0.1, \quad a_2 = 1.7$$

and determine for them the range of δ according to (18) and (35). In Tables

Part 1 of the boundary

$$\begin{cases} U = 2\sigma \log b, \\ V = 2(\sqrt{1-\sigma^2} - \sigma\, \overline{\text{arc}} \cos \sigma - \sqrt{b^2-\sigma^2} + \sigma\, \overline{\text{arc}} \cos \frac{\sigma}{b}) \end{cases}$$

$b = 0.1, \quad \sigma \in [0,0.1]$

σ	U	V
0	0.	1.800
0.02	-0.092	1.796
0.04	-0.184	1.785
0.06	-0.276	1.766
0.08	-0.368	1.738
0.1	-0.461	1.696

Part 2 of the boundary

$$\begin{cases} U = 2(\sigma \log \sigma - \sigma + b) \\ V = 2(\sqrt{1-\sigma^2} - \sigma\, \overline{\text{arc}} \cos \sigma) \end{cases}$$

$b = 0.1, \quad \sigma \in [0.1,1]$

σ	U	V	σ	U	V
0.1	-0.461	1.696	3.36	-1.256	1.000
0.12	-0.549	1.637	0.38	-1.295	0.952
0.14	-0.631	1.580	0.4	-1.333	0.906
0.16	-0.706	1.523	0.43	-1.386	0.837
0.18	-0.777	1.467	0.46	-1.434	0.770
0.2	-0.844	1.412	0.5	-1.493	0.685
0.22	-0.906	1.357	0.53	-1.533	0.623
0.24	-0.965	1.304	0.56	-1.569	0.563
0.26	-1.020	1.251	0.6	-1.613	0.487
0.28	-1.073	1.199	0.7	-1.699	0.315
3	-1.122	1.148	0.8	-1.757	0.170
3.32	-1.169	1.098	0.9	-1.790	0.060
3.34	-1.214	1.049	1	-1.8	0

Table 1.

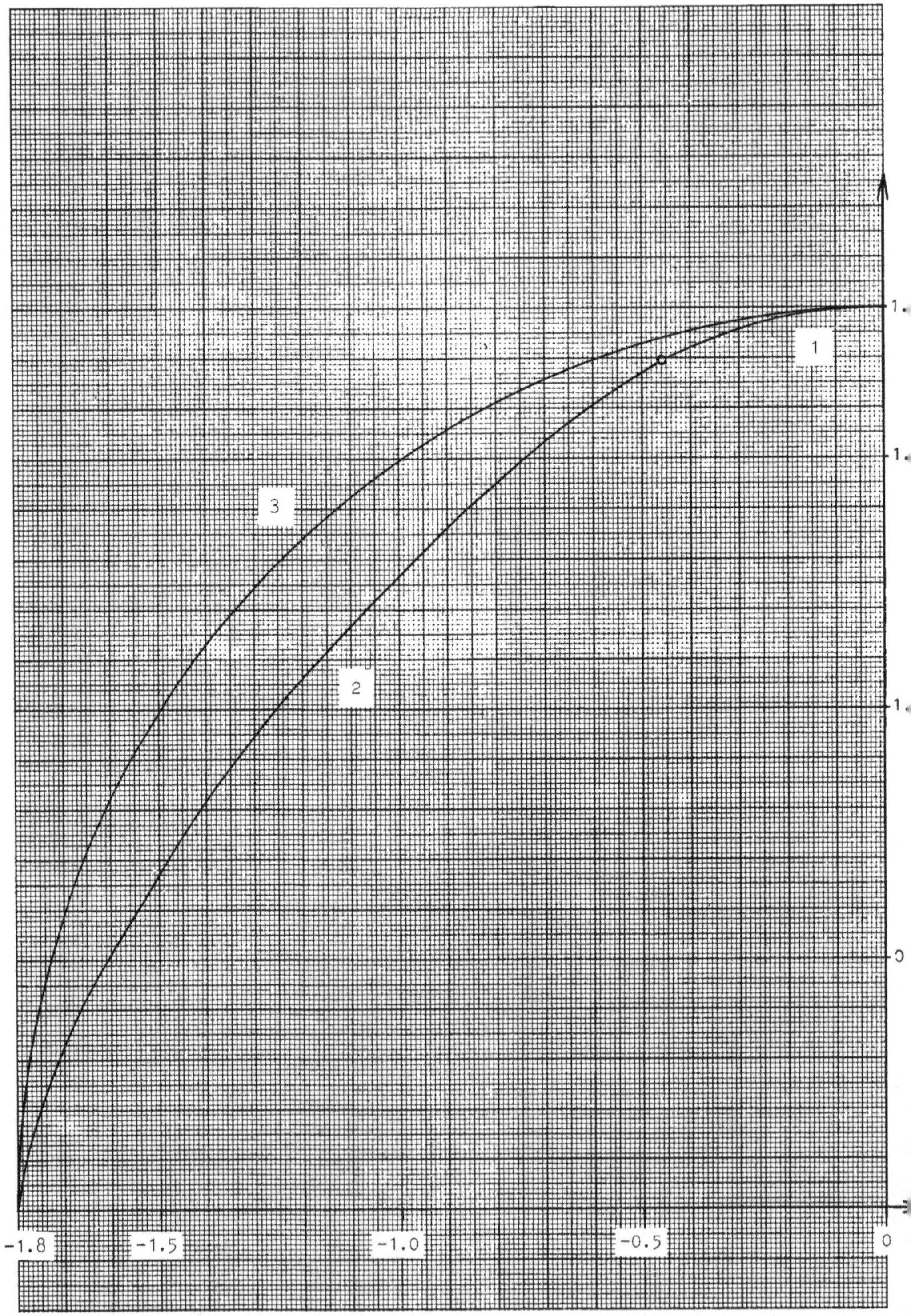

Figure 2.

The range of δ for $b = 0.1$, $a_2 = 1.5$

$$\begin{cases} v \in [\bar{v},\pi]; \quad \bar{v} = \overline{\text{arc}}\cos \dfrac{2b \log b}{|a_2|}, \\ \sigma \log \sigma - \sigma + b + \frac{1}{2}|a_2||\cos v| = 0, \\ \delta_1 = 2|a_2| \cos v \cdot \sigma + (1 - b^2 + 2(\sigma - b)^2) \cos 2v, \\ \delta_2 = -[2|a_2| \sin v \cdot \sigma + (1 - b^2 + 2(\sigma - b)^2) \sin 2v], \\ E(\sigma) = \sqrt{1 - \sigma^2} - \sigma\, \overline{\text{arc}} \cos \sigma - \frac{1}{2}|a_2||\sin v| \geqq 0. \end{cases}$$

v	σ	δ_1	δ_2	E
1.882˙847	0.1	-0.895˙477	0.293˙014	0.134
2	0.138˙405	-0.821˙825	0.373˙914	0.110
2.1	0.174˙206	-0.754˙600	0.421˙329	0.094
2.2	0.212˙341	-0.686˙905	0.451˙075	0.083
2.3	0.252˙268	-0.620˙471	0.465˙480	0.077
2.5	0.334˙537	-0.492˙004	0.454˙198	0.082
2.7	0.412˙813	-0.367˙081	0.386˙986	0.118
2.9	0.474˙476	-0.257˙067	0.249˙706	0.190
π	0.504˙979	-0.196˙921	0	0.337

Table 2.

The range of δ for $b = 0.1$, $a_2 = 1.7$

v	σ	δ_1	δ_2	E
1.845˙116	0.1	-0.936˙806	0.189˙025	0.030
1.9	0.120˙172	-0.915˙793	0.219˙594	0.014
1.95	0.139˙694	-0.896˙781	0.241˙838	0.000˙7
1.952˙808	0.140˙822	-0.895˙733	0.242˙938	0.000˙0
2	0.160˙279	-0.878˙637	0.259˙213	-0.012
2.1	0.204˙525	-0.847˙133	0.281˙642	-0.034
2.2	0.252˙645	-0.824˙099	0.291.940	-0.052
2.3	0.304˙297	-0.809˙728	0.295˙188	-0.065
2.4	0.359˙013	-0.801˙731	0.295˙364	-0.073
2.5	0.416˙127	-0.795˙960	0.294˙261	-0.074
2.6	0.474˙659	-0.787˙520	0.290˙703	-0.069
2.7	0.533˙127	-0.772˙268	0.280˙294	-0.055
2.8	0.589˙216	-0.748˙540	0.256˙026	-0.031
2.888˙550	0.634˙031	-0.722˙274	0.216˙721	0.000˙0
2.9	0.639˙308	-0.718˙744	0.210˙174	0.005
2.95	0.660˙498	-0.703˙649	0.177˙420	0.028
π	0.700˙920	-0.670˙918	0	0.157

$R = 0.362˙444, \quad \delta^{0} = -0.627˙556$

Table 3.

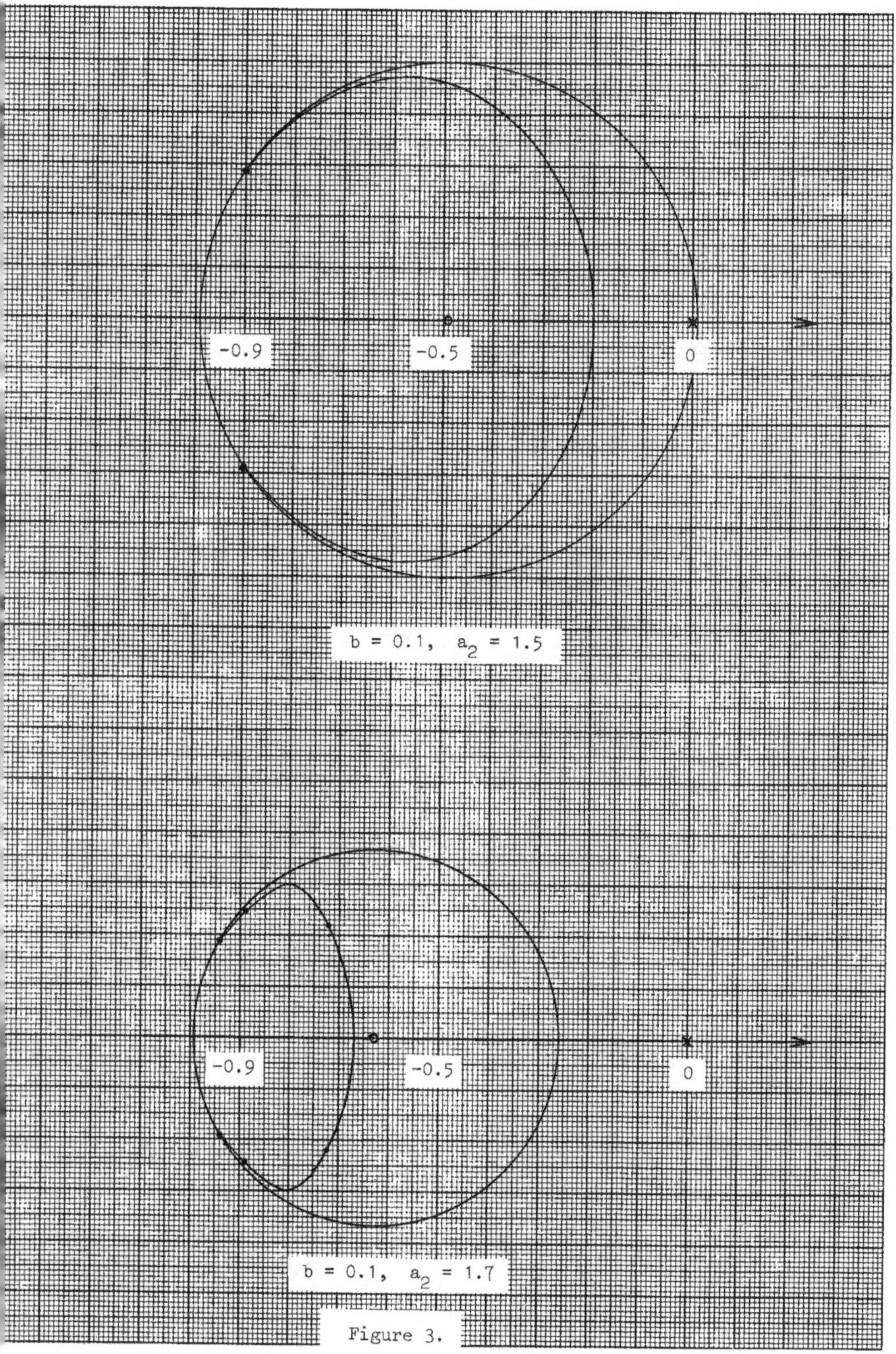

Figure 3.

2 and 3 there are the numerical values and in Figure 3 the corresponding range of δ. Observe, that only in the former case the whole domain obtained for δ is sharp, as is seen from the sign of the control function E.

In [4] the range of δ in the equality case 1:2 is determined by aid of considerations differing from the above perfect square method. Only the estimation technique based on the use of perfect squares is extendable to the next case 1:1.

4. The boundary functions 1:1

In this case we assume

$$x_0 = -C = -(C_1 + iC_2)$$

and write by aid of

$$\sigma = \sigma(u) > u \geqq b:$$

$$u\kappa^2 + 2x_0\kappa = u\kappa^2 - 2C\kappa = (u - \sigma)\left(\kappa - \frac{C}{2(u - \sigma)}\right)^2 + \sigma\left(\kappa - \frac{C}{2\sigma}\right)^2 - \frac{C^2 u}{4\sigma(u - \sigma)};$$

$$(36) \quad \begin{cases} \delta - 2Ca_2 - \dfrac{C^2}{2}\displaystyle\int_b^1 \frac{u}{\sigma(u - \sigma)}\,du \\ = -2\displaystyle\int_b^1 \left[\underbrace{(u - \sigma)}_{<\,0}\left(\kappa - \frac{C}{2(u - \sigma)}\right)^2 + \sigma\left(\kappa - \frac{C}{2\sigma}\right)^2\right]du \\ = 2\displaystyle\int_b^1 A^2 du - 2\int_b^1 B^2 du; \\ A = \sqrt{\sigma - u}\left(\kappa - \dfrac{C}{2(u - \sigma)}\right) = X + iY, \\ B = \sqrt{\sigma}\left(\kappa - \dfrac{C}{2\sigma}\right) = H + iK. \end{cases}$$

Because

$$\operatorname{Re} A^2 = |A|^2 - 2Y^2, \quad \operatorname{Re} B^2 = -|B|^2 + 2H^2,$$

we obtain from (36):

$$\operatorname{Re}\left(\delta - 2Ca_2 - \frac{C^2}{2}\int_b^1 \frac{u}{\sigma(u-\sigma)}\,du\right)$$

$$= 2\int_b^1 |A|^2 du + 2\int_b^1 |B|^2 du - 4\int_b^1 Y^2 du - 4\int_b^1 H^2 du.$$

Further, we have

$$|A|^2 = \sigma - u + \frac{|C|^2}{4(\sigma - u)} + \operatorname{Re}(\overline{C}\kappa),$$

$$|B|^2 = \sigma + \frac{|C|^2}{4\sigma} - \operatorname{Re}(\overline{C}\kappa);$$

$$|A|^2 + |B|^2 = 2\sigma - u + \frac{2\sigma - u}{4\sigma(\sigma - u)}|C|^2.$$

Hence

$$\operatorname{Re}(\delta - 2Ca_2) + 4\int_b^1 Y^2 du + 4\int_b^1 H^2 du$$

$$= \int_b^1 \left(\frac{\operatorname{Re} C^2}{2}\cdot\frac{u}{\sigma(u-\sigma)} + 4\sigma - 2u + \frac{2\sigma - u}{2\sigma(\sigma - u)}|C|^2\right) du$$

$$= -(1 - b^2) + \int_b^1 \left(4\sigma + \frac{(2\sigma - u)(C_1^2 + C_2^2) - u(C_1^2 - C_2^2)}{2\sigma(\sigma - u)}\right) du;$$

$$(37)\quad \begin{cases} \operatorname{Re}(\delta - 2Ca_2) = -(1 - b^2) + \overbrace{\int_b^1 \left(4\sigma + \frac{C_1^2}{\sigma} + \frac{C_2^2}{\sigma - u}\right) du}^{F} \\ \qquad - 4\int_b^1 Y^2 du - 4\int_b^1 H^2 du; \\ Y = \sqrt{\sigma - u}\left(\sin\vartheta - \frac{C_2}{2(u - \sigma)}\right), \\ H = \sqrt{\sigma}\left(\cos\vartheta - \frac{C_1}{2\sigma}\right). \end{cases}$$

This is the initial identity in the present case and corresponding to (1) used in the previous cases.

Let us require, that the parameters C_1 and C_2 both are $\neq 0$. In this case the function $\sigma = \sigma(u) > u$ minimizing the integrand

$$4\sigma + \frac{C_1^2}{\sigma} + \frac{C_2^2}{\sigma - u},$$

and thus the integral F, is determined uniquely by the condition

$$(38) \qquad \frac{1}{4}\frac{\partial}{\partial\sigma}\left(4\sigma + \frac{C_1^2}{\sigma} + \frac{C_2^2}{\sigma - u}\right) = 1 - \frac{C_1^2}{4\sigma^2} - \frac{C_2^2}{4(\sigma - u)^2} = 0.$$

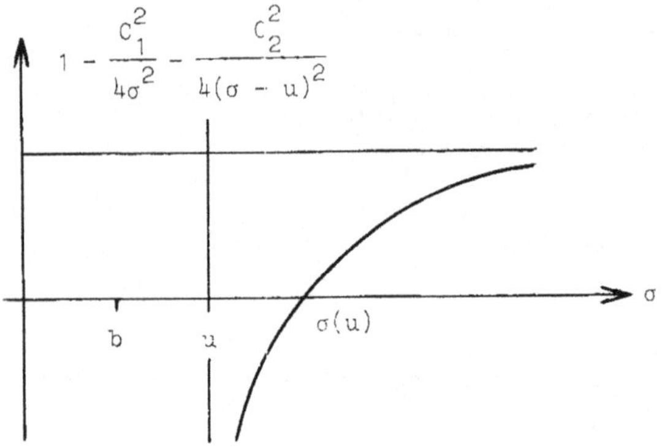

Figure 4.

The function $\sigma(u)$ obtained is monotonously increasing, because for $\sigma'(u)$ we have

$$\left(\frac{C_1^2}{2\sigma^3} + \frac{C_2^2}{2(\sigma - u)^3}\right)\sigma'(u) = \frac{C_2^2}{2(\sigma - u)^3}.$$

By aid of the connection $\sigma = \sigma(u)$ found we can introduce a function $\vartheta = \vartheta(u)$ for which the equality conditions

$$Y = H = 0$$

hold:

$$(39)\quad \begin{cases} \cos\vartheta = \dfrac{C_1}{2\sigma} \neq 0, \\[2ex] \sin\vartheta = \dfrac{C_2}{2(u-\sigma)} \neq 0. \end{cases}$$

The monotonous character of $\sigma = \sigma(u)$ implies the same property for $\vartheta = \vartheta(u)$ too. The monotonous inverse connection $u = u(\vartheta)$ therefore exists and is determined by (39) in the explicit form

$$(40)\qquad u = \frac{1}{2}\left(\frac{C_1}{\cos\vartheta} + \frac{C_2}{\sin\vartheta}\right).$$

Observe further that for $a_2 = U + iV$ determined by ϑ of (39) we have

$$\begin{cases} U = -2\displaystyle\int_b^1 \cos\vartheta \, du = -C_1 \int_b^1 \frac{du}{\sigma(u)}, \\[2ex] V = -2\displaystyle\int_b^1 \sin\vartheta \, du = C_2 \int_b^1 \frac{du}{\sigma(u) - u}. \end{cases}$$

Thus, for the function defined by (39)

$$UC_1 < 0, \quad VC_2 > 0.$$

For any generating function ϑ there holds

$$\operatorname{Re}(\delta - 2Ca_2) = -2\int_b^1 \operatorname{Re}(u\kappa^2 - 2C\kappa)du$$

$$= 4\int_b^1 \operatorname{Re}\left(-\frac{1}{2}u\kappa^2 + C\kappa\right)du$$

$$= 4\int_b^1 \left(-\frac{u}{2}\cos 2\vartheta + C_1\cos\vartheta - C_2\sin\vartheta\right)du.$$

This holds, of course, also for the function ϑ defined by (39) for which $Y = H = 0$. According to (37) we have for the S(b)-functions, in general,

$$\text{Re}\,(\delta - 2Ca_2) \leqq -(1 - b^2) + F.$$

Take now $\sigma(u)$ determined by (38). By aid of it we can define ϑ so that (39) holds. The right side of the previous inequality can now be rewritten by using this ϑ :

$$\text{(41)}\qquad \underset{f\in S(b)}{\text{Re}\,(\delta - 2Ca_2)} \leqq \underset{\vartheta\in(39)}{4\int_b^1} \left(-\frac{u}{2}\cos 2\vartheta + C_1\cos\vartheta - C_2\sin\vartheta\right)du$$

$$= -2\int_b^1 u\cos 2\vartheta\,du - 2C_1\underbrace{\left(-2\int_b^1 \cos\vartheta\,du\right)}_{U} + 2C_2\underbrace{\left(-2\int_b^1 \sin\vartheta\,du\right)}_{V}.$$

Equality here holds for f determined by ϑ from (39).

We can perform integrations on the right side of (41) by using the connection (40). This is done by introducing the terminal values for ϑ as follows:

$$\alpha = -\vartheta(1), \quad \omega = -\vartheta(b)$$

for which (40) gives

$$\text{(42)}\qquad \begin{cases} \dfrac{C_1}{\cos\alpha} - \dfrac{C_2}{\sin\alpha} = 2, \\[2ex] \dfrac{C_1}{\cos\omega} - \dfrac{C_2}{\sin\omega} = 2b. \end{cases}$$

Let us integrate the expressions of U and V. – Observe that (39) shows that $\cos\alpha$ and $\cos\omega$ as well as $\sin\alpha$ and $\sin\omega$ have the same sign. According to (40) we obtain:

$$du = \left(\frac{C_1}{2}\frac{\sin\vartheta}{\cos^2\vartheta} - \frac{C_2}{2}\frac{\cos\vartheta}{\sin^2\vartheta}\right)d\vartheta;$$

$$U = -2\int_b^1 \cos\vartheta\,du = C_2\int_{-\omega}^{-\alpha}\frac{\cos^2\vartheta}{\sin^2\vartheta}\,d\vartheta - C_1\int_{-\omega}^{-\alpha}\frac{\sin\vartheta}{\cos\vartheta}\,d\vartheta$$

$$= C_2(\cot \alpha - \cot \omega + \alpha - \omega) + C_1 \log \frac{\cos \alpha}{\cos \omega},$$

$$V = -2 \int_b^1 \sin \vartheta \, du = C_2 \int_{-\omega}^{-\alpha} \frac{\cos \vartheta}{\sin \vartheta} \, d\vartheta - C_1 \int_{-\omega}^{-\alpha} \frac{\sin^2 \vartheta}{\cos^2 \vartheta} \, d\vartheta$$

$$= C_2 \log \frac{\sin \alpha}{\sin \omega} + C_1(\tan \alpha - \tan \omega - \alpha + \omega).$$

If in these conditions U and V are explained to belong to an arbitrary $S(b)$-function, we obtain two necessary conditions for the existence of α and ω so that they form terminal values of a generating function ϑ making $Y = H = 0$. - Corresponding conditions are the first conditions (13) and (28). We shall return to the previous conditions $U = U(\alpha,\omega)$, $V = V(\alpha,\omega)$ later on.

Suppose now that α and ω are determined so that they belong to a given a_2 which is a coefficient in that $S(b)$-function for which the left side of (41) is written. This brings the expression $\operatorname{Re} (-2 \, Ca_2)$ on both sides of (41) and the inequality is reduced to the form

$$\operatorname{Re} \delta \leqq -2 \int_b^1 u \cos 2\vartheta \, du = 1 - b^2 - 4 \int_b^1 u \cos^2 \vartheta \, du$$

$\delta \in S(b) \quad \vartheta \in (39)$

$$= 1 - b^2 - 4 \int_{-\omega}^{-\alpha} \left(\frac{C_1}{2} \frac{1}{\cos \vartheta} + \frac{C_2}{2} \frac{1}{\sin \vartheta} \right) \cos^2 \vartheta \left(\frac{C_1}{2} \frac{\sin \vartheta}{\cos^2 \vartheta} - \frac{C_2}{2} \frac{\cos \vartheta}{\sin^2 \vartheta} \right) d\vartheta$$

$$= 1 - b^2 + C_1^2 \log \frac{\cos \alpha}{\cos \omega} - C_2^2 \log \frac{\sin \alpha}{\sin \omega} - \frac{C_2^2}{2} (\sin^{-2}\alpha - \sin^{-2}\omega)$$

$$+ C_1 C_2 (\cot \alpha - \cot \omega + 2(\alpha - \omega)) = G(\alpha,\omega).$$

Let us solve from (42) C_1 and C_2 in α and ω. The results found until now are collected.

<u>Theorem 5</u>. Let

(43) $\alpha = - \vartheta(1), \quad \omega = - \vartheta(b)$

be terminal values of a generating function to be determined by them.

Suppose that α and ω belong to the same quadrant and define the following numbers C_1, C_2 which are supposed to be $\neq 0$:

$$(44)\quad \begin{cases} C_1 = 2\dfrac{\sin\alpha - b\sin\omega}{\sin(\alpha-\omega)}\cos\alpha\cos\omega, \\ \\ C_2 = 2\dfrac{\cos\alpha - b\cos\omega}{\sin(\alpha-\omega)}\sin\alpha\sin\omega. \end{cases}$$

For a coefficient $a_2 = U + iV$ of a S(b)-function write the conditions

$$(45)\quad \begin{cases} U = C_1\log\dfrac{\cos\alpha}{\cos\omega} + C_2(\cot\alpha - \cot\omega + \alpha - \omega), \\ \\ V = C_1(\tan\alpha - \tan\omega - \alpha + \omega) + C_2\log\dfrac{\sin\alpha}{\sin\omega}. \end{cases}$$

Suppose that these equations determine α and ω in U and V so that $|\cos\omega| > |\cos\alpha|$. If this holds, then (39) together with (38) defines a generating function ϑ for which $|\cos\vartheta|$ is monotonously decreasing.

For the S(b)-function to which a_2 belongs the pair (α,ω) determines the inequality

$$(46)\quad \operatorname{Re}\delta \leqq 1 - b^2 + C_1^2\log\frac{\cos\alpha}{\cos\omega} - C_2^2\log\frac{\sin\alpha}{\sin\omega} - \frac{C_2^2}{2}(\sin^{-2}\alpha - \sin^{-2}\omega)$$

$$+ C_1C_2(\cot\alpha - \cot\omega + 2(\alpha-\omega)) = G(\alpha,\omega),$$

which is sharp for the generating function ϑ mentioned above. Thus, the existence of a pair (α,ω) for which $|\cos\omega| > |\cos\alpha|$ guarantees the sharpness of (46).

Next, consider the conditions $U = U(\alpha,\omega)$, $V = V(\alpha,\omega)$ defined by (44)-(45). Here (U,V) lies in E which is the complement of the domain defined by the boundary curves 1 and 2 of Figure 2. The pre-image of E in the $\alpha\omega$-plane is to be determined. Let us assume first that α and ω are restricted so that

$$(47)\qquad 0 < \omega < \alpha < \frac{\pi}{2}.$$

This restriction can be sharpened so that (α,ω) will lie in a triangle T and on ∂T the limit form of the inequality (46) holds. Moreover, $(\alpha,\omega) \in \overline{T}$

corresponds to points $(U,V) \in \overline{E}$. Consider the limit process involved.

1) $\omega \to 0$, $\alpha \neq 0$.

In this case $C_1 \to 2\cos\alpha$, $C_2 \to 0$. From (45) we obtain in the limit case

$$\begin{cases} U = 2\cos\alpha \log\cos\alpha - 2(\cos\alpha - b), \\ V = 2\sin\alpha - 2\alpha\cos\alpha \end{cases}$$

(46) gives in the limit case

$$\operatorname{Re}\delta \leqq 1 - b^2 + 4\cos^2\alpha \log\cos\alpha + 2(\cos\alpha - b)^2 - 4\cos\alpha\,(\cos\alpha - b).$$

When denoting

$$\cos\alpha = \sigma,$$

these formulae assume the form

$$U = 2(\sigma\log\sigma - \sigma + b),$$

$$V = 2(\sqrt{1-\sigma^2} - \sigma\,\overline{\text{arc}}\cos\sigma),$$

$$\operatorname{Re}\delta \leqq 1 - b^2 + 4\sigma^2\log\sigma + 2(\sigma - b)^2 - 4\sigma^2 + 4\sigma b$$

$$= 1 - b^2 + 2(\sigma - b)^2 + 2U\sigma.$$

Thus, we arrive at the limit case of the boundary function 1:2, dealt with in Theorem 3, conditions (25) and (28).

2) $\alpha \to \overline{\text{arc}}\cos(b\cos\omega)$.

Consider that part of the triangle (47) where

$$\cos\alpha \geqq b\cos\omega;$$

(48) $$\alpha \leqq \overline{\text{arc}}\cos(b\cos\omega).$$

In Figure 5 the triangle $T \ni (\alpha,\beta)$ is illustrated in the case $b = 0.1$.

In the equality case (48) we denote again

$$b\cos\omega = \cos\alpha = \sigma$$

and obtain from (44):

$$C_2 = 0,$$

$$C_1 = 2\,\frac{\sin\alpha - b\sin\omega}{\tan\alpha - \tan\omega} = 2\,\frac{\sqrt{1-\sigma^2} - b\sqrt{1-\frac{\sigma^2}{b^2}}}{\frac{\sqrt{1-\sigma^2}}{\sigma} - \sqrt{1-\frac{\sigma^2}{b^2}}\cdot\frac{b}{\sigma}} = 2\sigma.$$

Thus we see from (45)-(46)

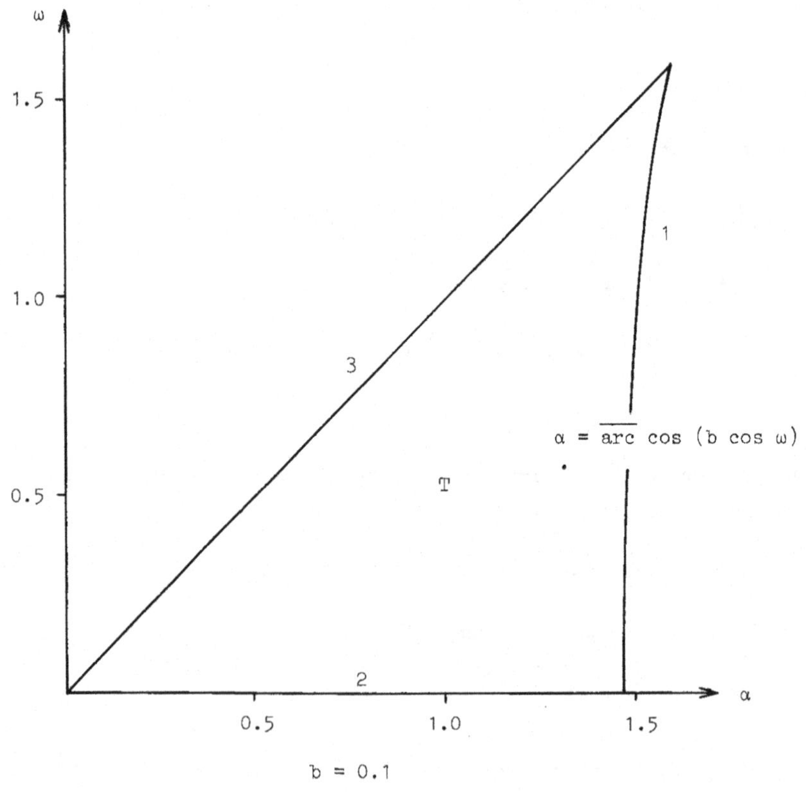

Figure 5.

$$U = 2\sigma \log b,$$

$$V = 2\left(\sqrt{1 - \sigma^2} - \sqrt{b^2 - \sigma^2} - \sigma\, \overline{\text{arc}} \cos \sigma + \sigma\, \overline{\text{arc}} \cos \frac{\sigma}{b}\right),$$

$$\text{Re}\ \delta \leqq 1 - b^2 + C_1^2 \log b = 1 - b^2 + \frac{U^2}{\log b},$$

i.e. we arrive at the inequality (9) of Theorem 1, in the limit case of 2:2 where one of the slits shrinks to a point, as is seen from (13).

3) $\omega \to \alpha$

This limit process is more complicated than the previous ones. For brevity, denote

$$\alpha - \omega = d; \quad \omega = \alpha - d$$

and observe that

$$\begin{cases} \log \dfrac{\cos \alpha}{\cos \omega} = -\tan \alpha \cdot d + O(d^2), \\[2ex] \log \dfrac{\sin \alpha}{\sin \omega} = \cot \alpha \cdot d + O(d^2), \\[2ex] \cot \alpha - \cot \omega + \alpha - \omega = -\cot^2 \alpha \cdot d + O(d^2), \\[2ex] \tan \alpha - \tan \omega - \alpha + \omega = \tan^2 \alpha \cdot d + O(d^2). \end{cases}$$

Thus we have:

$$C_1 \log \frac{\cos \alpha}{\cos \omega} = 2\, \frac{\sin \alpha - b \sin \omega}{\sin d} \cos \alpha \cos \omega(-\tan \alpha \cdot d + O(d^2))$$

$$\to -2(1 - b) \sin \alpha \tan \alpha \cos^2 \alpha = -2(1 - b) \sin^2 \alpha \cos \alpha,$$

$$C_2(\cot \alpha - \cot \omega + \alpha - \omega) = 2\, \frac{\cos \alpha - b \cos \omega}{\sin d} \sin \alpha \sin \omega(-\cot^2 \alpha \cdot d + O(d^2))$$

$$\to -2(1 - b) \cos \alpha \sin^2 \alpha \cot^2 \alpha = -2(1 - b) \cos^3 \alpha;$$

$$U \to -2(1 - b) \cos \alpha.$$

$$C_1(\tan\alpha - \tan\omega - \alpha + \omega) = 2\,\frac{\sin\alpha - b\sin\omega}{\sin d}\cos\alpha\cos\omega(\tan^2\alpha\cdot d + O(d^2))$$

$$\to 2(1-b)\sin\alpha\tan^2\alpha\cos^2\alpha = 2(1-b)\sin^3\alpha,$$

$$C_2\log\frac{\sin\alpha}{\sin\omega} = 2\,\frac{\cos\alpha - b\cos\omega}{\sin d}\sin\alpha\sin\omega(\cot\alpha\cdot d + O(d^2))$$

$$\to 2(1-b)\cos\alpha\sin^2\alpha\cot\alpha = 2(1-b)\sin\alpha\cos^2\alpha;$$

$$V \to 2(1-b)\sin\alpha.$$

In the limit case we thus have

$$a_2 = 2(1-b)e^{i(\pi-\alpha)},$$

i.e. the limiting equality function must belong to the radial-slit mapping. - In order to check that (46) gives in the limit case the corresponding inequality, we need expressions having two terms in their d-developments.

In order to find a development for the expression

$$(49)\qquad C_1^2\log\frac{\cos\alpha}{\cos\omega} - C_2^2\log\frac{\sin\alpha}{\sin\omega} - \frac{C_2^2}{2}(\sin^{-2}\alpha - \sin^{-2}\omega)$$

$$+ C_1C_2(\cot\alpha - \cot\omega + 2(\alpha-\omega))$$

we take out the factor $\dfrac{4}{\sin^2 d}$ and consider the second factor

$$I + II + III + IV;$$

$$\begin{cases} I = [(\sin\alpha - b\sin\omega)\cos\alpha\cos\omega]_1^2\log\dfrac{\cos\alpha}{\cos\omega}, \\[2ex] II = -[(\cos\alpha - b\cos\omega)\sin\alpha\sin\omega]_2^2\log\dfrac{\sin\alpha}{\sin\omega}, \\[2ex] III = -\dfrac{1}{2}(\sin^{-2}\alpha - \sin^{-2}\omega)[(\cos\alpha - b\cos\omega)\sin\alpha\sin\omega]_2^2, \\[2ex] IV = (\cot\alpha - \cot\omega + 2(\alpha+\omega))[\ ]_1\cdot[\ ]_2. \end{cases}$$

$$[\ ]_1 = (1-b)\sin\alpha\cos^2\alpha + \cos\alpha((1-b)\sin^2\alpha + b\cos^2\alpha)d + \dots,$$

$$[\ ]_1^2 = (1-b)^2\sin^2\alpha\cos^4\alpha + 2(1-b)\sin\alpha\cos^3\alpha((1-b)\sin^2\alpha + b\cos^2\alpha)d + \dots,$$

$$[\ ]_2 = (1-b)\cos\alpha\sin^2\alpha - \sin\alpha((1-b)\cos^2\alpha + b\sin^2\alpha)d + \dots,$$

$$[\ ]_2^2 = (1-b)^2\cos^2\alpha\sin^4\alpha - 2(1-b)\cos\alpha\sin^3\alpha((1-b)\cos^2\alpha + b\sin^2\alpha)d + \dots,$$

$$\log\frac{\cos\alpha}{\cos\omega} = -\tan\alpha\cdot d + \frac{1}{2\cos^2\alpha}d^2 + \dots,$$

$$\log\frac{\sin\alpha}{\sin\omega} = \cot\alpha\cdot d + \frac{1}{2\sin^2\alpha}d^2 + \dots,$$

$$\sin^{-2}\alpha - \sin^{-2}\omega = -2\frac{\cos\alpha}{\sin^3\alpha}d - \sin^{-2}\alpha(1+3\cot^2\alpha)d^2 + \dots,$$

$$\cot\alpha - \cot\omega + 2(\alpha-\omega) = \left(2 - \frac{1}{\sin^2\alpha}\right)d - \frac{\cos\alpha}{\sin^3\alpha}d^2 + \dots .$$

$$\mathrm{I} = -(1-b)^2\sin^3\alpha\cos^3\alpha\cdot d + \left[-2(1-b)\sin^2\alpha\cos^2\alpha((1-b)\sin^2\alpha + b\cos^2\alpha) + \frac{(1-b)^2}{2}\sin^2\alpha\cos^2\alpha\right]d^2 + \dots,$$

$$\mathrm{II} = -(1-b)^2\cos^3\alpha\sin^3\alpha\cdot d + \left[2(1-b)\cos^2\alpha\sin^2\alpha((1-b)\cos^2\alpha + b\sin^2\alpha) - \frac{(1-b)^2}{2}\cos^2\alpha\sin^2\alpha\right]d^2 + \dots,$$

$$\mathrm{III} = (1-b)^2\cos^3\alpha\sin\alpha\cdot d + \left[-2(1-b)\cos^2\alpha((1-b)\cos^2\alpha + b\sin^2\alpha) + \frac{(1-b)^2}{2}\cos^2\alpha(1+2\cos^2\alpha)\right]d^2 + \dots,$$

$$\mathrm{IV} = (1-b)^2(2\sin^3\alpha\cos^3\alpha - \sin\alpha\cos^3\alpha)d$$
$$+\left[2(1-b)(2b-1)\cos^2\alpha\sin^2\alpha\cos 2\alpha - (1-b)(2b-1)\cos^2\alpha\cos 2\alpha - (1-b)^2\cos^4\alpha\right]d^2 + \dots .$$

The sum I + II + III + IV has thus a development $Ad + Bd^2 + \dots$, where

$$A = -(1-b)^2\sin^3\alpha\cos^3\alpha - (1-b)^2\cos^3\alpha\sin^3\alpha + (1-b)^2\cos^3\alpha\sin\alpha$$
$$+ 2(1-b)^2\sin^3\alpha\cos^3\alpha - (1-b)^2\sin\alpha\cos^3\alpha = 0,$$

$$B = -2(1-b)^2\sin^4\alpha\,\cos^2\alpha - 2(1-b)b\,\sin^2\alpha\,\cos^4\alpha + 2(1-b)^2\cos^4\alpha\,\sin^2\alpha$$

$$+ 2(1-b)b\,\cos^2\alpha\,\sin^4\alpha - 2(1-b)^2\cos^4\alpha - 2(1-b)b\,\cos^2\alpha\,\sin^2\alpha$$

$$+ \frac{1}{2}\,(1-b)^2\cos^2\alpha + (1-b)(2b-1)\cos^2\alpha\,\cos 2\alpha(2\,\sin^2\alpha - 1)$$

$$= -2(1-b)^2(\cos^6\alpha + \cos^2\alpha\,\sin^4\alpha) - 4(1-b)b\,\cos^4\alpha\,\sin^2\alpha$$

$$+ \frac{1}{2}\,(1-b)^2\cos^2\alpha - (1-b)(2b-1)\cos^2\alpha\,\cos^2 2\alpha$$

$$= c_0 + c_1 b + c_2 b^2;$$

$$c_0 = -2\,\cos^6\alpha - 2\,\cos^2\alpha\,\sin^4\alpha + \frac{1}{2}\,\cos^2\alpha + \cos^2\alpha\,\cos^2 2\alpha$$

$$= \cos^2\alpha(\cos^2 2\alpha + \frac{1}{2} - 2\,\sin^4\alpha - 2\,\cos^4\alpha)$$

$$= \cos^2\alpha\,(\frac{1}{2} + (\cos^2\alpha - \sin^2\alpha)^2 - 2\,\sin^4\alpha - 2\,\cos^4\alpha)$$

$$= \cos^2\alpha(\frac{1}{2} - (\cos^2\alpha + \sin^2\alpha)^2)$$

$$= -\frac{1}{2}\,\cos^2\alpha,$$

$$c_1 = 4\,\cos^6\alpha + 4\,\cos^2\alpha\,\sin^4\alpha - 4\,\cos^4\alpha\,\sin^2\alpha - \cos^2\alpha - 3\,\cos^2\alpha(\cos^2\alpha - \sin^2\alpha)^2$$

$$= \cos^6\alpha + 2\,\cos^4\alpha\,\sin^2\alpha + \cos^2\alpha\,\sin^4\alpha - \cos^2\alpha$$

$$= \cos^4\alpha(\cos^2\alpha + \sin^2\alpha) + \cos^2\alpha\,\sin^2\alpha(\cos^2\alpha + \sin^2\alpha) - \cos^2\alpha$$

$$= \cos^2\alpha(\cos^2\alpha + \sin^2\alpha) - \cos^2\alpha$$

$$= 0,$$

$$c_2 = -2\,\cos^6\alpha - 2\,\cos^2\alpha\,\sin^4\alpha + 4\,\cos^4\alpha\,\sin^2\alpha + \frac{1}{2}\,\cos^2\alpha + 2\,\cos^2\alpha(\cos^2\alpha - \sin^2\alpha)^2$$

$$= \frac{1}{2}\,\cos^2\alpha.$$

The coefficient B is hence

$$B = -\frac{1}{2}\,(1 - b^2)\cos^2\alpha.$$

When multiplying $Bd^2 + \ldots$ by $\frac{4}{\sin^2 d}$ and by performing the limit process $d \to 0$ we obtain as the limit value of (49)

$-2(1 - b^2)\cos^2\alpha.$

Theorem 6. When passing by limit process to the boundary of the triangle T (Figure 5) the inequality (46) remains to hold and assumes the following forms on the parts 1, 2 and 3 of ∂T.

1: (9), equality for the limit case 2:2.

2: (25), equality for the limit case 1:2.

3: $\mathrm{Re}\,\delta = 1 - b^2 - 2(1-b^2)\cos^2\alpha$, holding for the radial slit mapping wit

$$a_2 = 2(1-b)e^{i(\pi-\alpha)}.$$

We may check directly the validity of the result obtained for the part 3 by taking $\cos\vartheta \equiv \cos\alpha$ in Löwner's formula:

$$\mathrm{Re}\,\delta = \mathrm{Re}\,(-2\int_b^1 u\kappa^2 du) = -2\int_b^1 u\cos 2\alpha du$$

$$= 1 - b^2 - 4\int_b^1 u\cos^2\alpha du = 1 - b^2 - 2(1 - b^2)\cos^2\alpha.$$

According to Theorem 5 the inequality (46) holds in T (Figure 5), because there the order of α and ω is correct. Similarly, the limit form of (46) holds on ∂T according to Theorem 6. Thus, the condition (46) is available in

$$\overline{T} = \{(\alpha,\omega) | 0 \leqq \omega \leqq \tfrac{\pi}{2}, \quad \omega \leqq \alpha \leqq \overline{\mathrm{arc}}\cos(b\cos\omega)\}.$$

In Figure 2 the curves 1 and 2 determine a domain where $\mathrm{Re}\,\delta$ has the sharp estimations (9) and (25). Call the complement of this domain E. We have to consider sharpness of the inequality (46) in E (Figure 6).

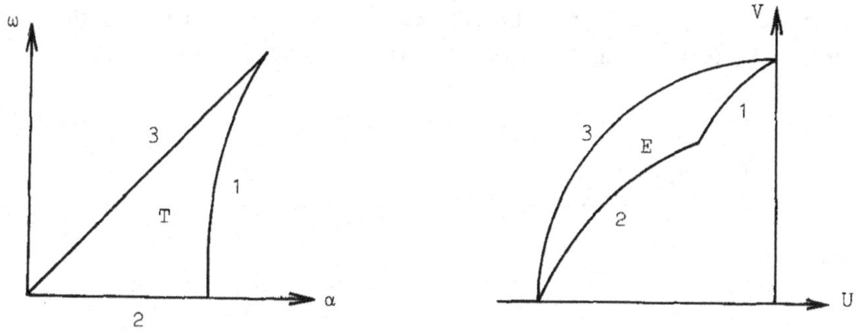

Figure 6.

The connection between T and E is defined by the equations (44)-(45), rewritten here:

$$(44)\text{-}(45)\quad \begin{cases} U = C_1 \log \dfrac{\cos\alpha}{\cos\omega} + C_2(\cot\alpha - \cot\omega + \alpha - \omega), \\[2ex] V = C_2 \log \dfrac{\sin\alpha}{\sin\omega} + C_1(\tan\alpha - \tan\omega - \alpha + \omega), \\[2ex] C_1 = 2\,\dfrac{\sin\alpha - b\sin\omega}{\sin(\alpha-\omega)}\cos\alpha\cos\omega, \\[2ex] C_2 = 2\,\dfrac{\cos\alpha - b\cos\omega}{\sin(\alpha-\omega)}\sin\alpha\sin\omega. \end{cases}$$

These conditions define the mapping

$$(50)\qquad g\colon \overline{T} \to \overline{E}, \quad g = (U,V),$$

with $U = U(\alpha,\omega)$, $V = V(\alpha,\omega)$ from (44)-(45). Observe, that this connection corresponds to the first conditions in (13) and (28). These bijective mappings between U and the parameter σ allowed us to utilize either σ or U when deriving the sharp inequalities (9) and (25) for $\operatorname{Re}\delta$. In the present case we are actually in the same situation:

Theorem 7. The mapping $g\colon \overline{T} \to \overline{E}$ defined by (44)-(45) is a bijective one. Thus, the right side of the inequality (46) can be explained to be a function of α,ω or U,V:

$$(51)\quad \operatorname{Re}\delta \leqq G(\alpha,\omega) = H(U,V).$$

The equations (44)-(45) allow only numerical checking of the above Theorem 7. In Figure 7 this correspondence is illustrated by drawing some curves in E which are images of the line segments α = constant in T.

Explicit expression for $H(U,V)$ can not be found. However, the expression δ in the equality case of (46) can be simplified as follows. From

$$\delta - Ca_2 = -2\int_b^1 (u\kappa^2 - C\kappa)\,du$$

we find that

$$\operatorname{Im}(\delta - Ca_2) = -2\int_b^1 (u \sin 2\vartheta - C_1 \sin\vartheta - C_2 \cos\vartheta)\,d\vartheta = 0,$$

because, according to (40),

$$(52) \qquad u \sin 2\vartheta - C_1 \sin\vartheta - C_2 \cos\vartheta = 0.$$

Further, (45) yields

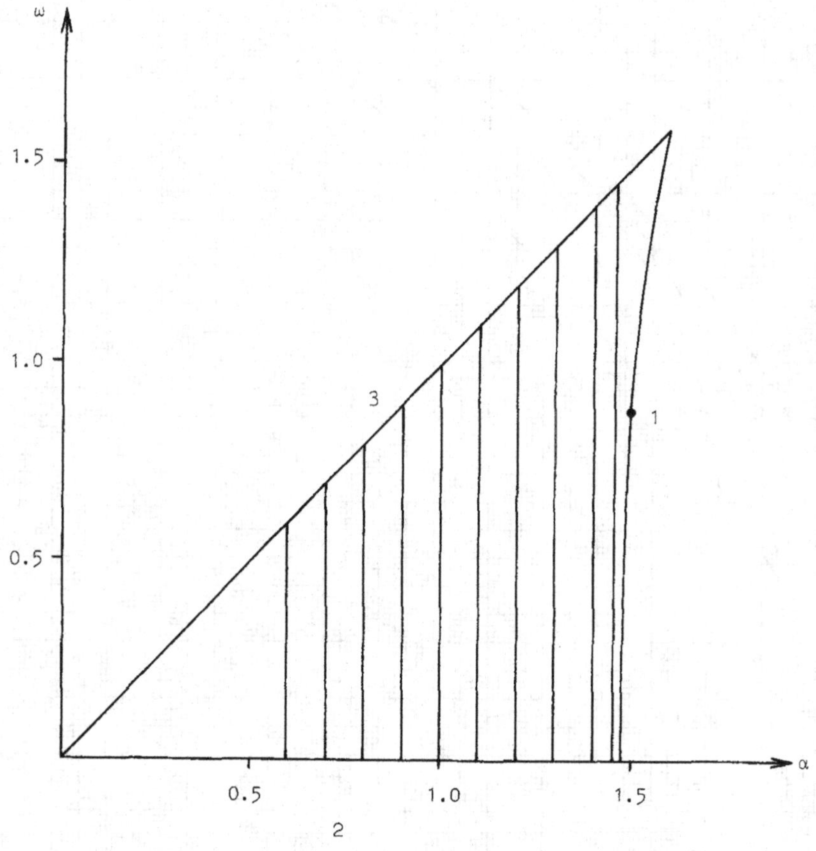

Figure 7 (part 1).

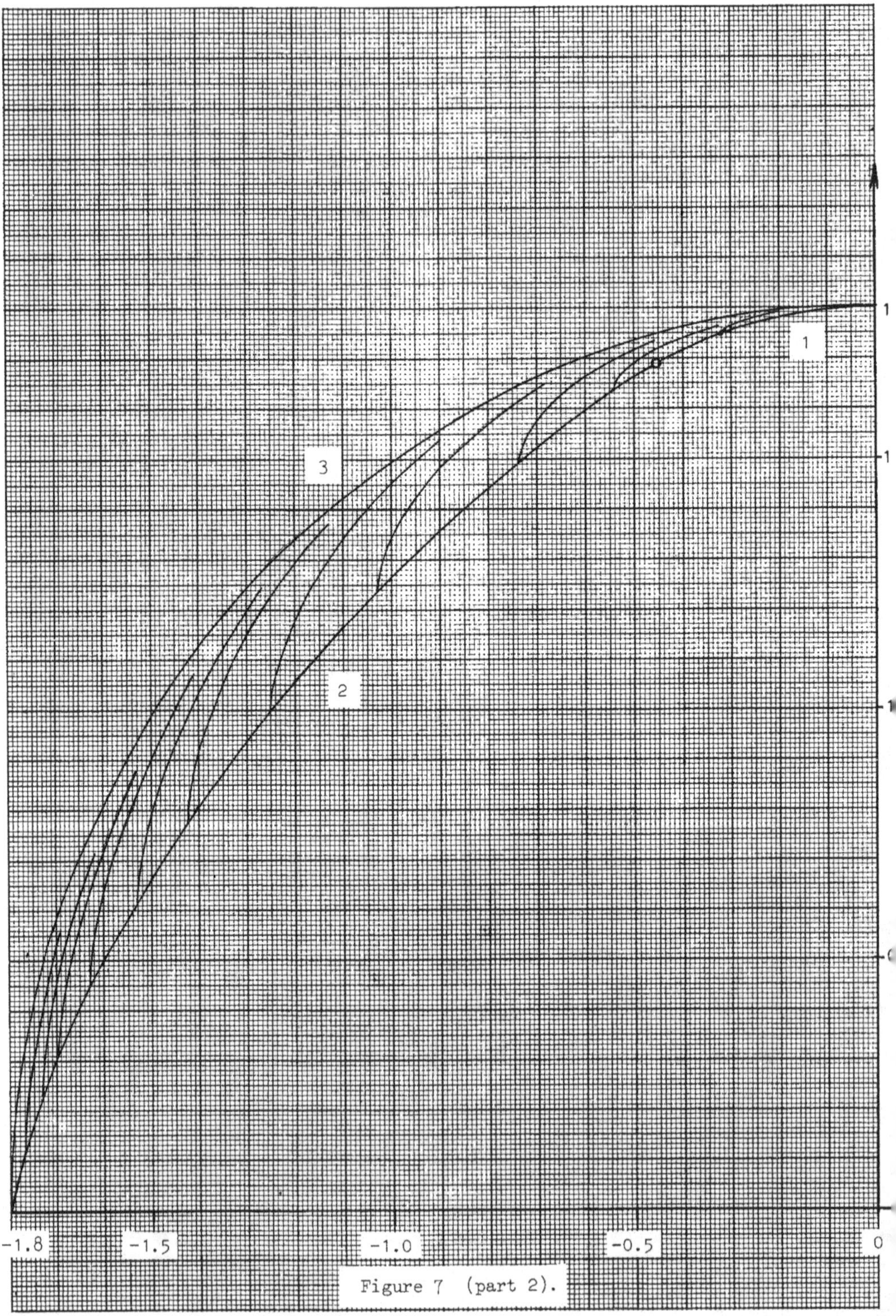

Figure 7 (part 2).

$$C_1U = C_1^2 \log \frac{\cos\alpha}{\cos\omega} + C_1C_2(\cot\alpha - \cot\omega + \alpha - \omega),$$

$$C_2V = C_2^2 \log \frac{\sin\alpha}{\sin\omega} + C_1C_2(\tan\alpha - \tan\omega - \alpha + \omega);$$

$$C_1U - C_2V = C_1^2 \log \frac{\cos\alpha}{\cos\omega} - C_2^2 \log \frac{\sin\alpha}{\sin\omega}$$

$$+ C_1C_2(\cot\alpha - \cot\omega + 2(\alpha - \omega)) + C_1C_2(\tan\omega - \tan\alpha)$$

and (46) can thus be rewritten in the form

$$\mathrm{Re}\,\delta \leqq 1 - b^2 + C_1C_2(\tan\alpha - \tan\omega) - \frac{C_2^2}{2}(\sin^{-2}\alpha - \sin^{-2}\omega) + C_1U - C_2V.$$

Because in the equality case

$$\mathrm{Im}\,\delta = \mathrm{Im}\,(Ca_2) = C_2U + C_1V,$$

and because

$$C_1U - C_2V + i(C_2U + C_1V) = U(C_1 + iC_2) + (iC_1 - C_2)V$$

$$= (C_1 + iC_2)(U + iV) = Ca_2,$$

we have the following expression (53) for δ belonging to the equality case.

Theorem 8. In the equality case of (46)

$$(53) \quad \delta = 1 - b^2 + Ca_2 + C_1C_2(\tan\alpha - \tan\omega) - \frac{C_2^2}{2}(\sin^{-2}\alpha - \sin^{-2}\omega),$$

where U, V, C_1 and C_2 are those in (44)-(45). Thus, (46) can be replaced by

$$(54) \quad \mathrm{Re}\,\delta \leqq 1 - b^2 + C_1U - C_2V + C_1C_2(\tan\alpha - \tan\omega) - \frac{C_2^2}{2}(\sin^{-2}\alpha - \sin^{-2}\omega).$$

Next, let us check if optimizing the parameters C_1, C_2 can give the condition (46) when one starts from the estimation (41), giving

$$(55) \qquad \mathrm{Re}\,\delta \leqq 2C_1U - 2C_2V + 4\int_b^1 \left(-\frac{u}{2}\cos 2\vartheta + C_1\cos\vartheta - C_2\sin\vartheta\right)du = M,$$

where (52) holds for ϑ . Thus, we determine α and ω so that

$$(56)\quad \begin{cases} \dfrac{\partial M}{\partial \alpha} = \dfrac{\partial M}{\partial C_1}\dfrac{\partial C_1}{\partial \alpha} + \dfrac{\partial M}{\partial C_2}\dfrac{\partial C_2}{\partial \alpha} = 0, \\[2ex] \dfrac{\partial M}{\partial \omega} = \dfrac{\partial M}{\partial C_1}\dfrac{\partial C_1}{\partial \omega} + \dfrac{\partial M}{\partial C_2}\dfrac{\partial C_2}{\partial \omega} = 0. \end{cases}$$

We omit the calculations needed to verify that

$$\frac{\partial(C_1,C_2)}{\partial(\alpha,\omega)} = \frac{\partial C_1}{\partial \alpha}\frac{\partial C_2}{\partial \omega} - \frac{\partial C_1}{\partial \omega}\frac{\partial C_2}{\partial \alpha} \neq 0.$$

This implies that (56) is equivalent to

$$(57)\quad \frac{\partial M}{\partial C_1} = \frac{\partial M}{\partial C_2} = 0.$$

The above derivatives can be obtained from (55), taking into account that (52) determines

$$\vartheta = \vartheta(u,C_1,C_2):$$

$$\frac{\partial M}{\partial C_1} = 2U + 4\int_b^1 [\underbrace{(u \sin 2\vartheta - C_1 \sin\vartheta - C_2 \cos\vartheta)}_{0}\frac{\partial \vartheta}{\partial C_1} + \cos\vartheta\,]du$$

$$= 2U + 4\int_b^1 \cos\vartheta \, du,$$

$$\frac{\partial M}{\partial C_2} = -2V - 4\int_b^1 \sin\vartheta \, du.$$

Hence, (57) yields

$$(58)\quad \begin{cases} U = -2\displaystyle\int_b^1 \cos\vartheta \, du, \\[2ex] V = -2\displaystyle\int_b^1 \sin\vartheta \, du. \end{cases}$$

As we have seen before, this, together with (52), leads to the conditions (45). - Hence, also in the present case optimizing the parameters C_ν leads to the same result which comes out when the parameters are determined so that (58) holds.

In order to reach other parts of the disc $|a_2| \leqq 2(1-b)$ besides the one lying in the second quadrant, we have to restrict α and ω correspondingly. This is achieved by reflecting the point (α,ω) with respect to the real and imaginary axes in the $\alpha\omega$-plane, as illustrated in Figure 8. As a result we obtain the following cases 1°-4° of Figure 8. The inequality (46) remains unchanged in all these cases.

Finally, we may again generalize the result by aid of the rotation $\tau^{-1}f(\tau z)$ which, when applied to Theorems 5 and 8, gives:

<u>Theorem 9</u>. In $S(b)$

$$(59)\quad \operatorname{Re}(\tau^2\delta) \leqq 1-b^2+C_1\tilde{U}-C_2\tilde{V}+C_1C_2(\tan\alpha-\tan\omega)-\frac{C_2^2}{2}(\sin^{-2}\alpha-\sin^{-2}\omega)$$

where $\tau = e^{iv}$ and $a_2 = U + iV$ determine the numbers (α,ω) so that

$$(60)\quad \begin{cases} \tilde{U} = \operatorname{Re}(\tau a_2) = \cos v\cdot U-\sin v\cdot V = C_1\log\dfrac{\cos\alpha}{\cos\omega}+C_2(\cot\alpha-\cot\omega+\alpha-\omega), \\ \tilde{V} = \operatorname{Im}(\tau a_2) = \sin v\cdot U+\cos v\cdot V = C_2\log\dfrac{\sin\alpha}{\cos\omega}+C_1(\tan\alpha-\tan\omega-\alpha+\omega); \\ C_1 = 2\,\dfrac{\sin\alpha - b\sin\omega}{\sin(\alpha-\omega)}\cos\alpha\cos\omega, \\ C_2 = 2\,\dfrac{\cos\alpha - b\cos\omega}{\sin(\alpha-\omega)}\sin\alpha\sin\omega. \end{cases}$$

The numbers τ and a_2 for which (59) is sharp must determine $(\alpha,\omega) \in \overline{T}$ and $(\tilde{U},\tilde{V}) \in \overline{E}$ according to Figure 8.

The number δ lies in the intersection of the half planes given by (60). The boundary curve of this intersection is given by

$$(61)\quad \delta = C\tau^{-1}a_2 + \tau^{-2}\left[1-b^2+C_1C_2(\tan\alpha-\tan\omega)-\frac{C_2^2}{2}(\sin^{-2}\alpha-\sin^{-2}\omega)\right]$$

$$= \delta_1 + i\delta_2.$$

As an example we may again consider the case

$1^o \quad 0 < \omega < \alpha < \frac{\pi}{2}; \quad U < 0, \ V > 0.$

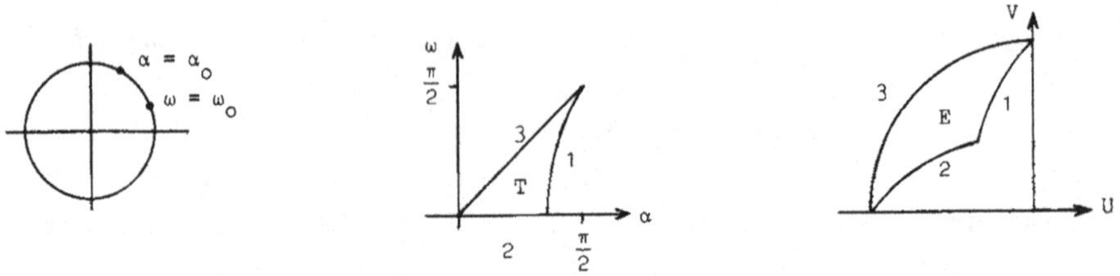

$2^o \quad -\frac{\pi}{2} < \alpha < \omega < 0; \quad U < 0, \ V < 0.$

$\alpha = -\alpha_o, \ \omega = -\omega_o; \quad C_1 = C_1^o, \ C_2 = -C_2^o; \quad U = U^o, \ V = -V^o.$

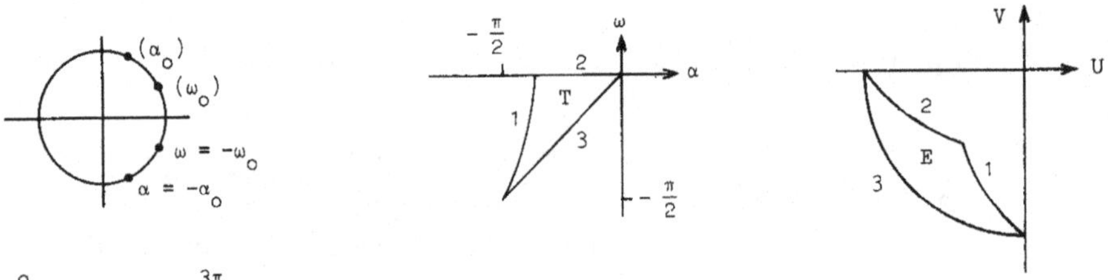

$3^o \quad \pi < \omega < \alpha < \frac{3\pi}{2}; \quad U > 0, \ V < 0.$

$\alpha = \pi + \alpha_o, \ \omega = \pi + \omega_o; \quad C_1 = -C_1^o, \ C_2 = -C_2^o; \quad U = -U^o, \ V = -V^o.$

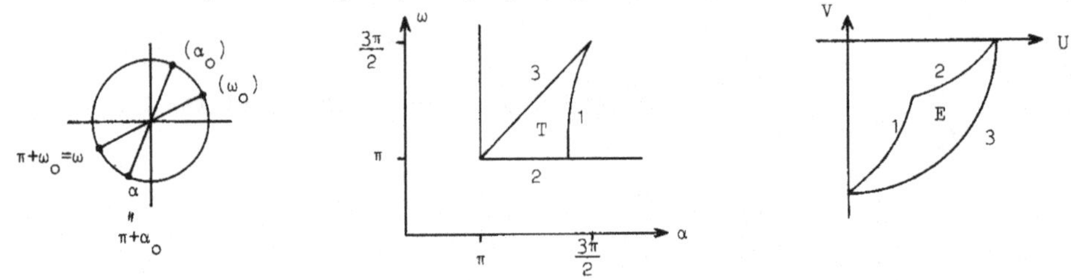

$4^o \quad -\frac{3\pi}{2} < \alpha < \omega < -\pi; \quad U > 0, \ V > 0.$

$\alpha = -(\pi + \alpha_o), \ \omega = -(\pi + \omega_o); \quad C_1 = -C_1^o, \ C_2 = C_2^o; \quad U = -U^o, \ V = V^o.$

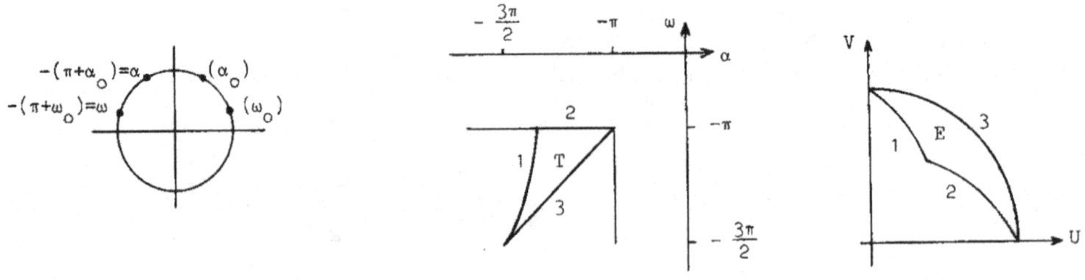

Figure 8.

$$U = a_2 = |a_2| \geqq 0,$$

for which

$$(62) \quad \begin{cases} \widetilde{U} = \cos v \cdot U, \qquad \widetilde{V} = \sin v \cdot U, \\ \delta = \delta_1 + i\delta_2 = [C_1 \cos v + C_2 \sin v + i(C_2 \cos v - C_1 \sin v)]U \\ \quad + e^{-i2v} [1 - b^2 + C_1C_2(\tan \alpha - \tan \omega) - \frac{C_2^2}{2} (\sin^{-2}\alpha - \sin^{-2}\omega)]. \end{cases}$$

Take $b = 0.1$, $a_2 = 1.7$ and $v \in [\frac{\pi}{2},\pi]$ in which case $(\widetilde{U},\widetilde{V})$ belongs to the second quadrant, case 1° in Figure 8. In Table 4 there are the numerical values of (α,ω), $(\widetilde{U},\widetilde{V})$ and (δ_1,δ_2) of the boundary points. Figure 9 illustrates the corresponding part of the boundary curve belonging to the cross section of the coefficient body.

$b = 0.1, \quad a_2 = 1.7, \quad v \in [\frac{\pi}{2},\pi]$

v	α	ω	$\widetilde{U}$	$\widetilde{V}$	δ_1	δ_2
1.952˙808	1.429˙505	0.000˙000	-0.633˙733	1.577˙458	-0.895˙733	0.242˙938
2	1.407˙070	0.072˙054	-0.707˙450	1.545˙806	-0.881˙713	0.256˙320
2.1	1.352˙212	0.166˙751	-0.858˙238	1.467˙456	-0.858˙457	0.272˙890
2.2	1.291˙787	0.193˙313	-1.000˙452	1.374˙444	-0.839˙520	0.281˙268
2.3	1.228˙466	0.186˙382	-1.136˙669	1.267˙699	-0.822˙874	0.284˙899
2.4	1.164˙111	0.161˙951	-1.253˙569	1.148˙287	-0.807˙396	0.285˙104
2.5	1.100˙315	0.128˙360	-1.361˙944	1.017˙403	-0.792˙269	0.282˙222
2.6	1.038˙618	0.090˙835	-1.456˙711	0.876˙352	-0.776˙781	0.275˙860
2.7	0.980˙587	0.053˙425	-1.536˙923	0.726˙546	-0.760˙215	0.264˙737
2.8	0.927˙584	0.020˙239	-1.601˙778	0.569˙480	-0.741˙720	0.245˙951
2.86	0.898˙144	0.004˙833	-1.633˙044	0.472˙406	-0.729˙110	0.228˙298
2.888˙551	0.884˙041	0.000˙000	-1.645˙864	0.425˙595	-0.722˙274	0.216˙721

Table 4.

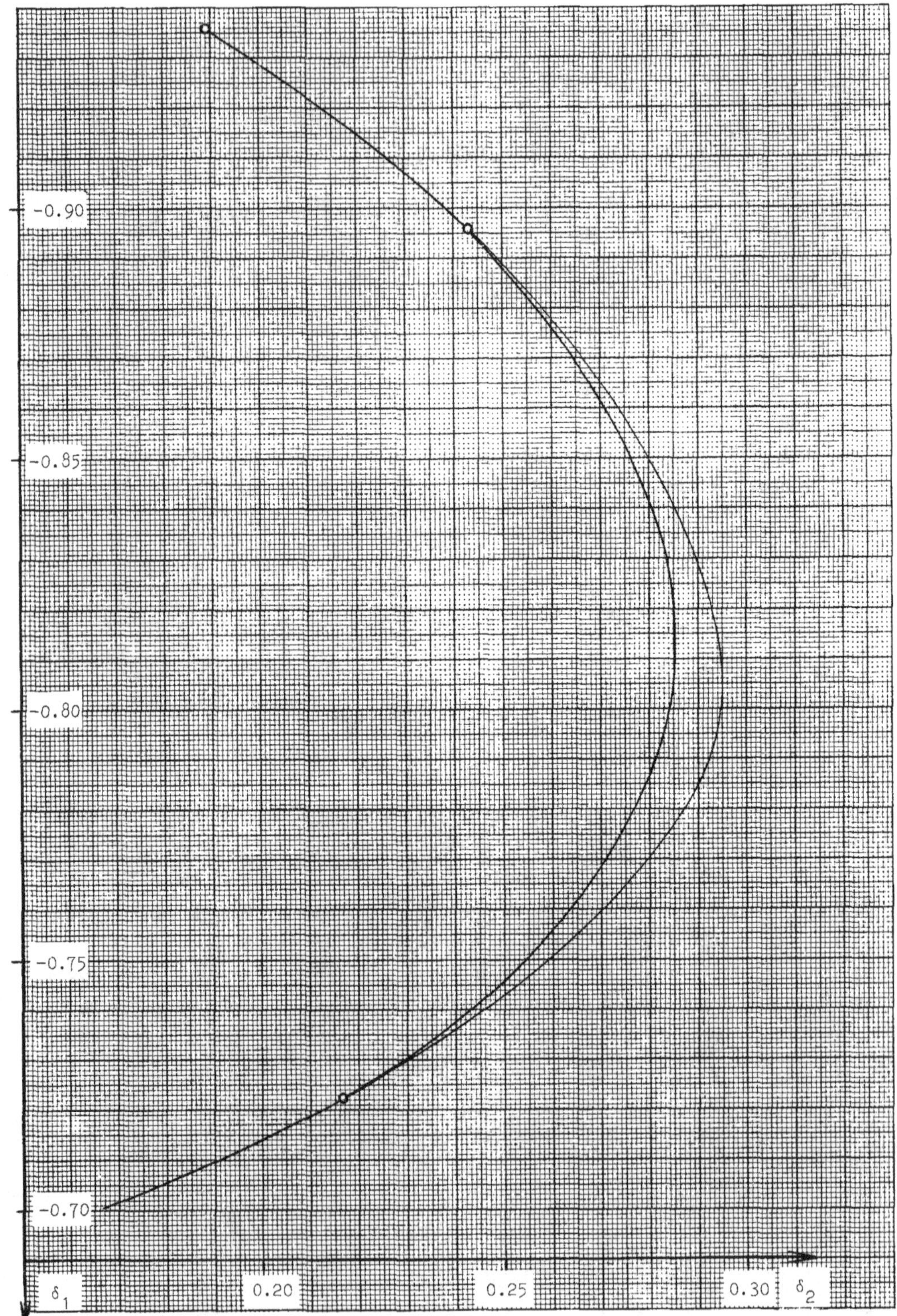

Figure 9.

2 § The Linear Combination $a_3 + \lambda a_2$ in $S_R(b)$

1. $\min (a_3 + \lambda a_2)$

Let $S_R(b) \subset S(b)$ denote that subclass of $S(b)$ where all the coefficients $a_2, a_3, \ldots$ are real. In [1], pp. 229, 230, 234 the following sharp estimations, determining the first coefficient body (a_2, a_3) in $S_R(b)$, are given:

$$a_3 \geqq -(1 - b^2) + a_2^2 \tag{1}$$

$$a_3 \leqq \begin{cases} 1 - b^2 + (1 + \frac{1}{\log b})a_2^2, & |a_2| \leqq 2b\ |\log b|. \\ 1 - b^2 + a_2^2 - 2\sigma|a_2| + 2(\sigma - b)^2, & |a_2| \geqq 2b\ |\log b|. \end{cases} \tag{2}$$

Here $\sigma \in [e^{-1}, 1]$ is the root of the equation

$$\sigma \log \sigma - \sigma + b + \frac{|a_2|}{2} = 0. \tag{3}$$

For the combination $a_3 + \lambda a_2$, $\lambda \in R$, these estimations give sharp bounds in a_2. We shall determine min and max for $a_3 + \lambda a_2$ in $S_R(b)$.

In order to find $\min (a_3 + \lambda a_2)$ we have to minimize the lower limit given by (1):

$$\begin{aligned} a_3 + \lambda a_2 &\geqq -(1 - b^2) + a_2^2 + \lambda a_2 \\ &= -(1 - b^2) - \frac{\lambda^2}{4} + (a_2 + \frac{\lambda}{2})^2 = A(a_2), \quad |a_2| \leqq 2(1 - b). \end{aligned} \tag{4}$$

The location of the point $a_2 = -\frac{\lambda}{2}$ with respect to the interval $[-2(1-b), 2(1-b)]$ determines the following three alternatives:

$$|-\frac{\lambda}{2}| \leqq 2(1 - b); \quad \min A = A(-\frac{\lambda}{2}),$$

$$-\frac{\lambda}{2} \leqq -2(1 - b); \quad \min A = A(-2(1 - b)),$$

$$-\frac{\lambda}{2} \geqq 2(1 - b); \quad \min A = A(2(1 - b)).$$

<u>Theorem 1</u>. In $S_R(b)$

$$(5)\qquad \min\,(a_3 + \lambda a_2) = \begin{cases} 3 - 8b + 5b^2 + \lambda 2(1 - b) & \text{for } \lambda \leqq -4(1 - b), \\ -(1 - b^2) - \frac{\lambda^2}{4} & \text{for } -4(1 - b) \leqq \lambda \leqq 4(1 - b), \\ 3 - 8b + 5b^2 - \lambda 2(1 - b) & \text{for } \lambda \geqq 4(1 - b). \end{cases}$$

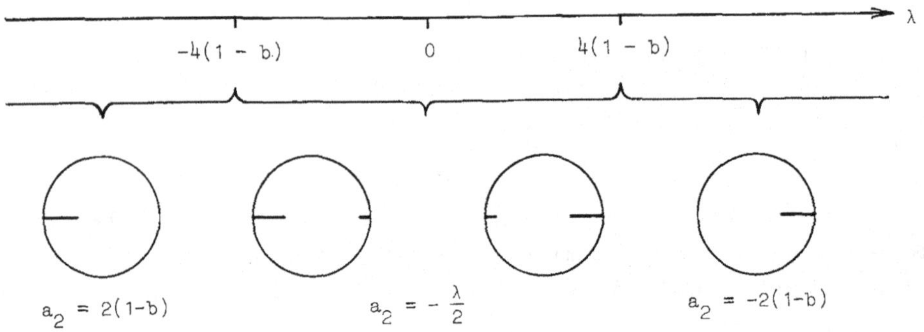

Figure 10.

2. $\max\,(a_3 + \lambda a_2)$ for $e^{-1} \leqq b < 1$

Now we have to maximize

$$(6)\qquad a_3 + \lambda a_2 \leqq \begin{cases} 1 - b^2 + (1 + \frac{1}{\log b})a_2^2 + \lambda a_2, & |a_2| \leqq 2b\,|\log b|, \\ 1 - b^2 + a_2^2 - 2\sigma|a_2| + \lambda a_2 + 2(\sigma - b)^2, & |a_2| \geqq 2b\,|\log b| \end{cases}$$

If $\lambda a_2 \leqq 0$ one can perform the rotation $\tau^{-1} f(\tau z)$, $\tau = -1$, which changes the sign of a_2 but does not affect a_3. Thus, maximum is reached in the case where

$$\lambda a_2 = |\lambda a_2| \geqq 0.$$

We may restrict ourselves to the case $\lambda \geqq 0$ for which only the cases $a_2 \geqq 0$ need to be studied. For $\lambda \leqq 0$, $a_2 \leqq 0$ we see by using the rotation mentioned, that maximum is found by replacing in the first case λ and a_2 by $|\lambda|$ and $|a_2|$.

Suppose first, that $e^{-1} < b < 1$ and consider separately the two intervals included in (6), namely

$$1^\circ.\quad 0 \leqq a_2 \leqq 2b\,|\log b|, \qquad 2^\circ.\quad 2b\,|\log b| \leqq a_2 \leqq 2(1-b).$$

In these cases the upper bounds of (6) are to be maximized.

1°. $0 \leqq a_2 \leqq 2b\,|\log b|$.

Denote the quantity to be maximized

$$B(a_2) = 1 - b^2 + \rho a_2^2 + \lambda a_2 = 1 - b^2 - \frac{\lambda^2}{4\rho} + \rho\left(a_2 + \frac{\lambda}{2\rho}\right)^2,$$

$$\rho = 1 + \frac{1}{\log b} < 0.$$

There are the following two alternatives 1) and 2) according to the location of the point $-\frac{\lambda}{2\rho}$.

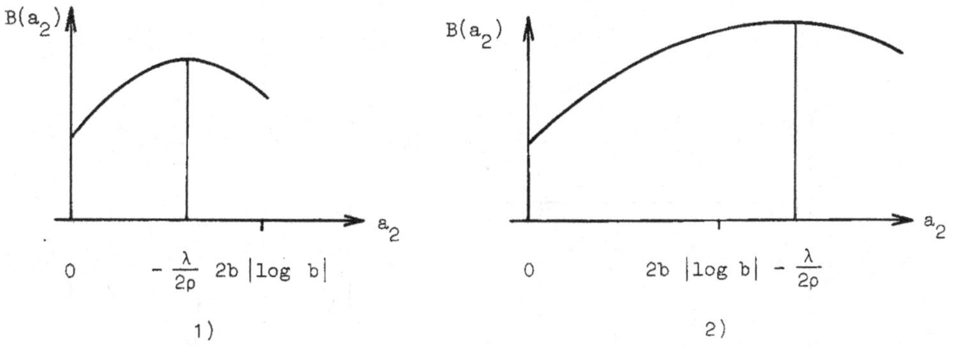

Figure 11.

1) $0 \leqq -\frac{\lambda}{2\rho} \leqq 2b\,|\log b| \iff 0 \leqq \lambda \leqq 4b(1 + \log b)$.

$$\max B = B(-\frac{\lambda}{2\rho}) = 1 - b^2 - \frac{\lambda^2}{4}\frac{\log b}{1 + \log b}.$$

2) $-\frac{\lambda}{2\rho} \geqq 2b\,|\log b| \iff \lambda \geqq 4b(1 + \log b)$.

$$\max B = B(2b\,|\log b|) = 1 - b^2 + 4b^2 \log b(1 + \log b) - 2\lambda b \log b.$$

2°. $2b\,|\log b| \leqq a_2 \leqq 2(1 - b)$.

Denote now the upper bound from (6)

$$\tilde{C}(a_2) = 1 - b^2 + a_2^2 - 2\sigma a_2 + \lambda a_2 + 2(\sigma - b)^2,$$

where

(7) $$\sigma \log \sigma - \sigma + b + \frac{a_2}{2} = 0.$$

Figure 12 illustrates the fact that (7) determines $\sigma = \sigma(a_2)$ on the interval 2°, where

$$-2(1 - b) \leqq -a_2 \leqq -2b\,|\log b|.$$

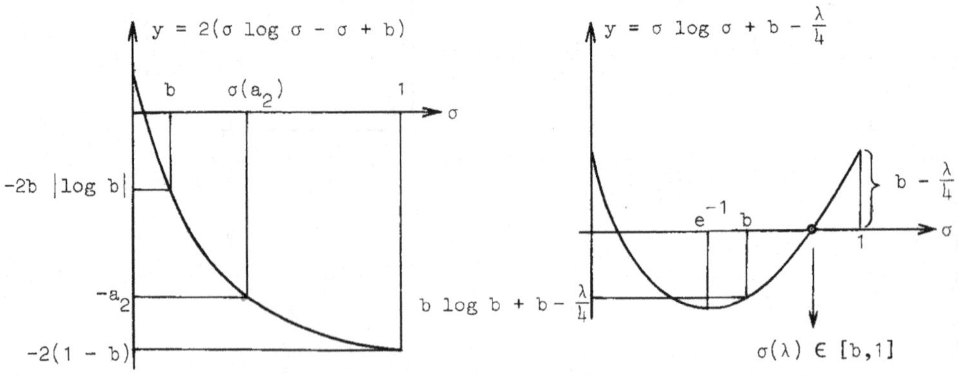

Figure 12.

Thus, from (7) we obtain

$$\sigma(a_2) \in [b,1] \quad \text{for} \quad a_2 \in [2b\,|\log b|,\ 2(1-b)].$$

Let us in $\tilde{C}(a_2)$ use the simpler inverse connection

$$a_2 = a_2(\sigma) = 2(\sigma - \sigma \log \sigma - b)$$

which gives, when denoting $\tilde{C}(a_2(\sigma)) = C(\sigma)$:

$$C(\sigma) = 1-b^2+4(\sigma-\sigma \log \sigma-b)^2+2(\lambda-2\sigma)(\sigma-\sigma \log \sigma-b)+2(\sigma-b)^2,$$

where $\sigma \in [b,1]$. For this function there holds

$$C'(\sigma) = 8 \log \sigma \,(\sigma \log \sigma + b - \tfrac{\lambda}{4}).$$

As illustrated in Figure 12, the equation

$$\sigma \log \sigma + b - \frac{\lambda}{4} = 0$$

has a root $\sigma(\lambda) \in [b,1]$ so far as

$$b - \frac{\lambda}{4} \geqq 0 \quad \text{and} \quad b \log b + b - \frac{\lambda}{4} \leqq 0$$

i.e.

$$0 < 4b\,(1 + \log b) \leqq \lambda \leqq 4b.$$

From the sign of $C'(\sigma)$ we deduce that if $\sigma(\lambda) \in (b,1)$, then $C(\sigma(\lambda))$ is the maximum of C. If $b \log b + b - \frac{\lambda}{4} \geqq 0$, then $\sigma \log \sigma + b - \frac{\lambda}{4} \geqq 0$, $C'(\sigma) \leqq 0$ and thus $C(b)$ is the maximum. Finally, if $b - \frac{\lambda}{4} \leqq 0$ then $\sigma \log \sigma + b - \frac{\lambda}{4} \leqq 0$, $C'(\sigma) \geqq 0$ and $C(1)$ is the maximum. Thus, in the case 2° we have

$$\max C = \begin{cases} C(b) & \text{for} \quad 0 \leqq \lambda \leqq 4b\,(1 + \log b), \\ C(\sigma(\lambda)) & \text{for} \quad 4b\,(1 + \log b) \leqq \lambda \leqq 4b, \\ C(1) & \text{for} \quad \lambda \geqq 4b. \end{cases}$$

We have to compare the two alternatives given by 1° and 2°. On the interval

$$0 \leqq \lambda \leqq 4b\ (1 + \log b)$$

there holds

$$B(-\frac{\lambda}{2\rho}) \geqq C(b)$$

with the equality only for $\lambda = 4b\ (1 + \log b)$. This is due to the fact that

$$B(-\frac{\lambda}{2\rho}) = 1-b^2-\frac{\lambda^2}{4}\frac{\log b}{1+\log b} \geqq 1-b^2+4b^2 \log b(1+\log b)-2\lambda b \log b = C(b)$$

$\Longleftrightarrow$

$$[\lambda - 4b\ (1 + \log b)]^2 \geqq 0.$$

On the interval

$$4b\ (1 + \log b) \leqq \lambda \leqq 4b$$

there holds

$$B(2b\ |\log b|) = C(b) \leqq C(\sigma(\lambda)).$$

This is seen as follows. In the present case $C'(\sigma)$ was found to be positive for $b \leqq \sigma \leqq \sigma(\lambda)$. Thus the above inequality holds, with the equality only for $\sigma(\lambda) = b$, which belongs to the limit case $\lambda = 4b\ (1 + \log b)$.

For the values

$$\lambda \geqq 4b$$

there holds the inequality

$$B(2b\ |\log b|) = C(b) \leqq C(1).$$

This follows from the fact that $C'(\sigma) \geqq 0$ for $\lambda \geqq 4b$.

Finally, when substituting $b = e^{-1}$ in the expression of B we can directly check that the above results remain to hold in the form where the interval $[0, 4b\ (1 + \log b)]$ vanishes.

Theorem 2. In $S_R(b)$ and $b \in [e^{-1},1]$ there holds for $\lambda \geqq 0$:

$$(8)\quad \max\,(a_3+\lambda a_2) = \begin{cases} B(-\frac{\lambda}{2\rho}) = 1 - b^2 - \frac{\lambda^2}{4}\frac{\log b}{1+\log b} & \text{for } 0 \leqq \lambda \leqq 4b(1+\log b), \\ C(\sigma(\lambda)) = 1-b^2+\lambda(\sigma(\lambda)-\frac{\lambda}{4})+2(\sigma(\lambda)-b)^2 & \text{for } 4b(1+\log b) \leqq \lambda \leqq 4b, \\ C(1) = 3 - 8b + 5b^2 + \lambda\cdot 2(1-b) & \text{for } \lambda \geqq 4b. \end{cases}$$

Here $\sigma(\lambda) \in [b,1]$ is the root of the equation

$$(9)\quad \sigma \log \sigma + b - \frac{\lambda}{4} = 0.$$

For $\lambda \leqq 0$ the maximum is obtained from the above formulae by using $|\lambda|$ in them instead of λ.

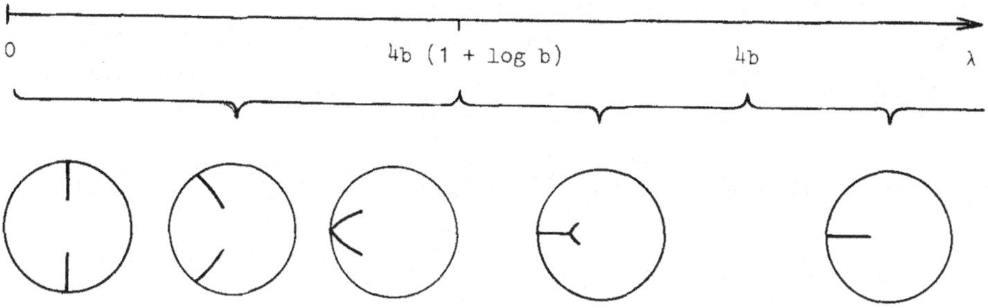

Figure 13.

Observe that for $b = e^{-1}$ the first interval $[0,4b(1+\log b)]$ shrinks to the point $\lambda = 0$. All the types of the extremal domains valid in this interval remain to hold at the point $\lambda = 0$, $b = e^{-1}$.

3. max $(a_3 + \lambda a_2)$ for $0 < b < e^{-1}$

In the present case

$$\rho = 1 + \frac{1}{\log b} > 0.$$

Consider first the case $\lambda \geqq 0$.

Take the case 1° where $0 \leqq a_2 \leqq 2b|\log b|$ and the quantity to be maximized is B. Because

$$-\frac{\lambda}{2\rho} \leqq 0$$

we see that

$$\max B = B(2b|\log b|).$$

Hence, $\max B$ is reached at the endpoint of the interval 2°: $2b|\log b| \leqq a_2 \leqq 2(1-b)$.

In order to find $\max(a_3 + \lambda a_2)$ we thus have to maximize the quantity C defined on the interval 2°. Consider $C'(\sigma)$ and the equation (9). Because $b < e^{-1}$ the inequality

$$b \log b + b - \frac{\lambda}{4} < 0$$

holds automatically and thus the only condition for the existence of the root $\sigma(\lambda) \in [e^{-1}, 1]$ is $b - \frac{\lambda}{4} \geqq 0 \Longleftrightarrow 0 \leqq \lambda \leqq 4b$. $C'(\sigma)$ changes its sign from positive to negative at $\sigma(\lambda)$, which implies that

$$\max C = C(\sigma(\lambda)) \quad \text{for} \quad 0 \leqq \lambda \leqq 4b.$$

If $\lambda \geqq 4b$, then $C'(\sigma) \geqq 0$ and hence

$$\max C = C(1) \quad \text{for} \quad \lambda \geqq 4b.$$

<u>Theorem 3</u>. In $S_R(b)$ and $b \in (0, e^{-1})$ there holds for $\lambda \geqq 0$:

$$(10)\quad \max(a_3+\lambda a_2) = \begin{cases} C(\sigma(\lambda)) = 1-b^2+\lambda(\sigma(\lambda)-\frac{\lambda}{4})+2(\sigma(\lambda)-b)^2 & \text{for } 0 \leqq \lambda \leqq 4b, \\ C(1) = 3-8b+5b^2+\lambda\cdot 2(1-b) & \text{for } \lambda \geqq 4b. \end{cases}$$

Here $\sigma(\lambda) \in [e^{-1}, 1]$ is the root of the equation

$$\sigma \log \sigma + b - \frac{\lambda}{4} = 0.$$

For $\lambda \leqq 0$ the reasoning of the Theorem 2 remains to hold.

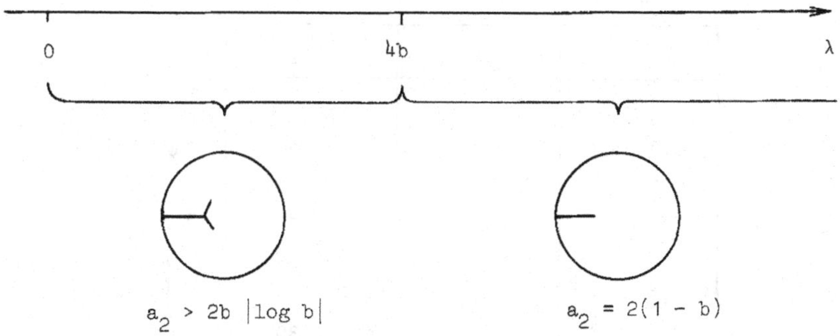

Figure 14.

Table 5 and Figure 15 belong to an example where $b = 0.5 > e^{-1}$ and where the numerical values of $\max\ (a_3 + \lambda a_2)$ are determined for some values of λ.

$S_R(b)$; max $(a_3 + \lambda a_2)$

$$e^{-1} < b = 0.5 \Rightarrow \begin{cases} 4b(1 + \log b) = 0.613\cdot706 \\ 4b = 2 \end{cases}$$

λ	σ	max $(a_3+\lambda a_2)$	
0		0.750·000	
0.1		0.755·647	
0.2		0.772·589	
0.3		0.800·825	$B(-\frac{\lambda}{2\rho})$
0.4		0.840·356	
0.5		0.891·181	
0.6		0.953·300	
0.613·706	0.5	0.962·694	
0.7	0.559·281	1.026·025	
0.8	0.612·993	1.105·929	
0.9	0.658·698	1.190·698	
1	0.699·491	1.279·084	
1.1	0.736·868	1.370·268	
1.2	0.771·691	1.463·661	
1.3	0.804·508	1.558·811	$C(\sigma)$
1.4	0.835·695	1.655·356	
1.5	0.865·523	1.752·999	
1.6	0.894·194	1.851·488	
1.7	0.921·865	1.950·610	
1.8	0.948·659	2.050·176	
1.9	0.974·677	2.150·021	
2	1	2.25	
2.1		2.35	
2.2		2.45	
2.3		2.55	
2.4		2.65	
2.5		2.75	
2.6		2.85	$C(1)$
2.7		2.95	
2.8		3.05	
2.9		3.15	
3		3.25	

Table 5.

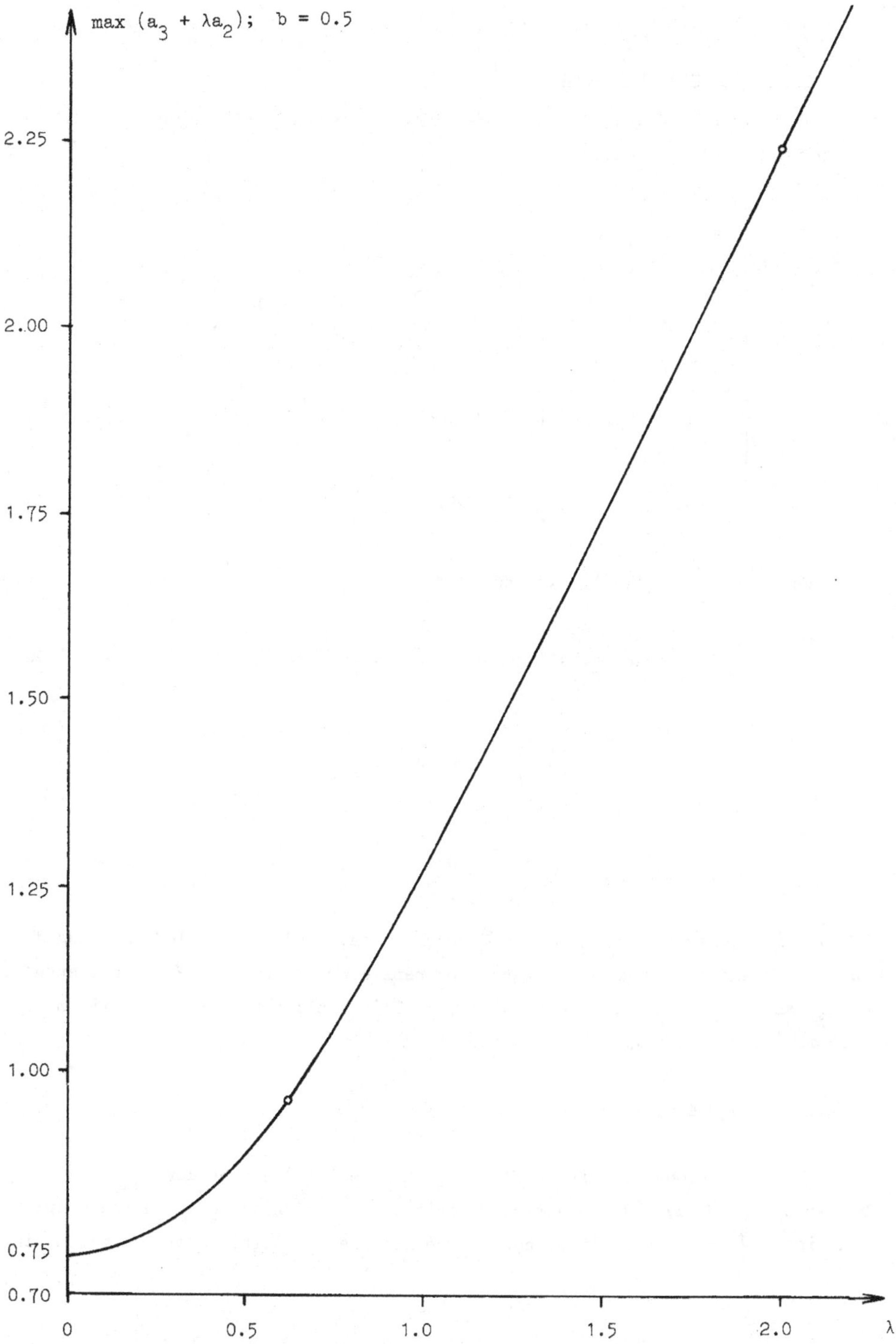
max $(a_3 + \lambda a_2)$; b = 0.5
2.25
2.00
1.75
1.50
1.25
1.00
0.75
0.70
0
0.5
1.0
1.5
2.0
λ

Figure 15.

3 § Re $(a_3 + \lambda a_2)$ in S(b)

1. Preliminary Remarks

For complex values of λ we are going to use Löwner's expressions in the combination $a_3 + \lambda a_2$:

$$\begin{cases} a_2 = -2\int_b^1 \kappa du, \\ \delta = a_3 - a_2^2 = -2\int_b^1 u\kappa^2 du; \quad \kappa = \kappa(u) = e^{i\vartheta(u)}, \\ U = \operatorname{Re} a_2 = -2\int_b^1 \cos\vartheta\, du, \\ V = \operatorname{Im} a_2 = -2\int_b^1 \sin\vartheta\, du. \end{cases}$$

When denoting $\lambda = \lambda_1 + i\lambda_2$ we thus have

$$(1) \qquad \operatorname{Re}(a_3 + \lambda a_2) = 4\left(\int_b^1 \cos\vartheta\, du\right)^2 - 4\left(\int_b^1 \sin\vartheta\, du\right)^2 - 2\int_b^1 u\cos 2\vartheta\, du$$

$$-2\lambda_1\int_b^1 \cos\vartheta\, du + 2\lambda_2\int_b^1 \sin\vartheta\, du$$

$$= \operatorname{Re} a_3 + \lambda_1 U - \lambda_2 V.$$

If ϑ is replaced in this by $-\vartheta$ then Re a_3 and U are left unchanged but V changes the sign. Similarly, by replacing ϑ by $\pi - \vartheta$ we preserve Re a_3 and V and change the sign of U. This implies that in the maximum case of (1) necessarily

$$(2) \qquad \lambda_1 U \geqq 0, \quad \lambda_2 V \leqq 0.$$

The variational condition for the functional (1) can be derived either by replacing $\vartheta(u)$ by the varied function $\vartheta(u) + \varepsilon\eta(u)$ or by using step-functions ϑ and the limit process. The necessary condition for ϑ maximizing

(1), i.e. the variational condition, assumes the form obtainable by aid of a formal differentiation $\frac{\partial}{\partial \vartheta}$:

$$8\int_b^1 \cos\vartheta\, du\cdot(-\sin\vartheta) - 8\int_b^1 \sin\vartheta\, du\cdot\cos\vartheta + 4u\sin 2\vartheta$$

$$+ 2\lambda_1 \sin\vartheta + 2\lambda_2 \cos\vartheta = 0$$

$\Rightarrow$

$$(3)\qquad \begin{cases} C_1 \sin\vartheta + C_2 \cos\vartheta = 2u \sin\vartheta \cos\vartheta\,, \\ C_1 = -(U + \frac{\lambda_1}{2}), \\ C_2 = -(V + \frac{\lambda_2}{2}). \end{cases}$$

Observe that this is the condition (40)/V.1.4 under a special requirement for the parameter $C = C_1 + iC_2$:

$$(4)\qquad C = -(a_2 + \frac{\lambda}{2}).$$

This connection with the coefficient body (a_2,a_3) will be used later on.

When considering the expression (1) we see that the part of it where u exists can be written in the form

$$-\int_b^1 u\cos^2\vartheta\, du \quad \text{or} \quad \int_b^1 u\sin^2\vartheta\, du.$$

By using step-functions ϑ we easily read out from these expressions that in the maximum case $|\cos\vartheta|$ is necessarily decreasing and $|\sin\vartheta|$ increasing in the interval $[b,1]$.

2. The Extremal Case 2:2

We are going to utilize the corresponding identity which was used in V.1 for studying the coefficient body (a_2,a_3):

$$\begin{cases} \delta + x_0 a_2 + \frac{x_0^2}{2} \log b = -2 \int_b^1 A^2 du, \\ A = \sqrt{u}(\kappa + \frac{x_0}{2u}) = X + iY. \end{cases}$$

This gives us

$$\operatorname{Re}\,(\delta + x_0 a_2 + \frac{x_0^2}{2} \log b) = -2 \int_b^1 \operatorname{Re} A^2 du$$

$$= 2 \int_b^1 |A|^2 du - 4 \int_b^1 X^2 du$$

$$= 1 - b^2 - \frac{|x_0|^2}{2} \log b - \operatorname{Re}\,(\overline{x}_0 a_2) - 4 \int_b^1 X^2 du.$$

The condition (4) suggested by the variational condition appears to be useful here. Therefore, we substitute in the above identity

$$x_0 = -C = -C_1 - iC_2 = a_2 + \frac{1}{2}\lambda;$$

$$C_1 = -(U + \frac{\lambda_1}{2}), \quad C_2 = -(V + \frac{\lambda_2}{2}).$$

Thus

$$\operatorname{Re}\,(\delta + \underbrace{(x_0 + \overline{x}_0)}_{-2C_1} a_2) = 1 - b^2 - \frac{1}{2}(\underbrace{|x_0|^2 + \operatorname{Re} x_0^2}_{2C_1^2}) \log b - 4 \int_b^1 X^2 du$$

$\Rightarrow$

$$\operatorname{Re}\,(a_3 + \lambda a_2) = \operatorname{Re}\,(a_2^2) + \operatorname{Re}\,(\lambda a_2) + 2C_1 \operatorname{Re} a_2 + 1 - b^2 - C_1^2 \log b - 4 \int_b^1 X^2 du$$

$$= U^2 - V^2 + \lambda_1 U - \lambda_2 V - 2(U + \frac{\lambda_1}{2})U + 1 - b^2 - (U + \frac{\lambda_1}{2})^2 \log b - 4 \int_b^1 X^2 du.$$

Write this by aid of two perfect squares in the form

$$(5) \quad \operatorname{Re}\,(a_3 + \lambda a_2) = 1 - b^2 + \frac{\lambda_2^2}{4} - (V + \frac{\lambda_2}{2})^2 - (1 + \log b)[U + \frac{\lambda_1 \log b}{2(1 + \log b)}]^2$$

$$- \frac{\lambda_1^2}{4} \frac{\log b}{1 + \log b} - 4 \int_b^1 X^2 du; \qquad X = \sqrt{u}(\cos \vartheta + \frac{\operatorname{Re} x_0}{2u}).$$

Let us consider first the interval $e^{-1} < b < 1$ where $1 + \log b > 0$. There we have

$$\begin{cases} \operatorname{Re}(a_3 + \lambda a_2) \leqq 1 - b^2 - \frac{\lambda_1^2}{4\rho} + \frac{\lambda_2^2}{4}; . \\ \rho = 1 + \frac{1}{\log b} < 0. \end{cases}$$

Observe that this inequality is true for all numbers λ_1, λ_2, but the equality can be reached if and only if

(6) $$U = - \frac{\lambda_1 \log b}{2(1 + \log b)},$$

(7) $$V = - \frac{\lambda_2}{2},$$

(8) $$\cos \vartheta = \frac{C_1}{2u} = \frac{\sigma}{u}; \quad \sigma = - \frac{\operatorname{Re} x_0}{2} = \frac{C_1}{2}.$$

We have to check the existence of such ϑ for which all the conditions (6)-(8) hold simultaneously.

Assume first that C_1 is limited so that

$$0 \leqq \sigma = \frac{C_1}{2} \leqq b.$$

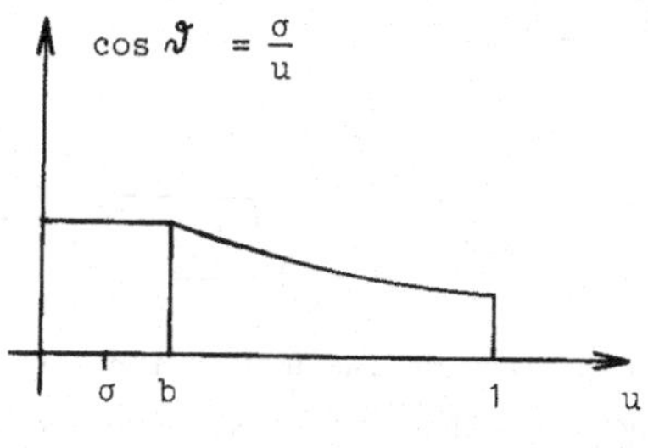

Figure 16.

The function ϑ defined by (8) gives for U:

$$U = -C_1 - \frac{\lambda_1}{2} = -2\sigma - \frac{\lambda_1}{2} = -2\int_b^1 \cos\vartheta \, du = -2\int_b^1 \frac{\sigma}{u}\, du = 2\sigma \log b \leqq 0$$

$\Rightarrow$

$$0 \leqq \sigma = \frac{C_1}{2} = -\frac{\lambda_1}{4(1+\log b)}; \quad \lambda_1 \leqq 0;$$

(9) $$U = -\frac{\lambda_1 \log b}{2(1+\log b)} \leqq 0.$$

Thus we see that the two conditions (6) and (8) fit together and hold simultaneously provided that $\lambda_1 \leqq 0$ and

$$\sigma = \frac{|\lambda_1|}{4(1+\log b)} \leqq b.$$

For ϑ we have

(10) $$\cos\vartheta = \frac{\sigma}{u}, \quad \sin\vartheta = \begin{cases} \sqrt{1-\frac{\sigma^2}{u^2}}, & b \leqq u \leqq c, \\ -\sqrt{1-\frac{\sigma^2}{u^2}}, & c \leqq u \leqq 1. \end{cases}$$

Here c is to be determined so that, according to (7),

$$-\frac{\lambda_2}{2} = V = -2\int_b^1 \sin\vartheta \, du = -2\int_b^c \sqrt{1-\frac{\sigma^2}{u^2}}\, du - 2\int_c^1 -\sqrt{1-\frac{\sigma^2}{u^2}}\, du.$$

This is possible as far as for λ_2 holds

(11) $$\left|\frac{\lambda_2}{4}\right| = \left|\frac{V}{2}\right| \leqq \int_b^1 \sqrt{1-\frac{\sigma^2}{u^2}}\, du$$

$$= \sqrt{1-\sigma^2} - \sigma \,\overline{\text{arc}}\cos\sigma - \sqrt{b^2-\sigma^2} + \sigma\,\overline{\text{arc}}\cos\frac{\sigma}{b}.$$

The above equality conditions define a $S(b)$-function of the type 2:2 and equality in (11) belongs to such limit case where one of the slits shrinks to a point.

Next we assume that $C_1 \leqq 0$ and denote now

$$0 \leqq \sigma = -\frac{C_1}{2} \leqq b.$$

In the equality case (8) we have thus

$$\cos\vartheta = \frac{C_1}{2u} = -\frac{\sigma}{u}.$$

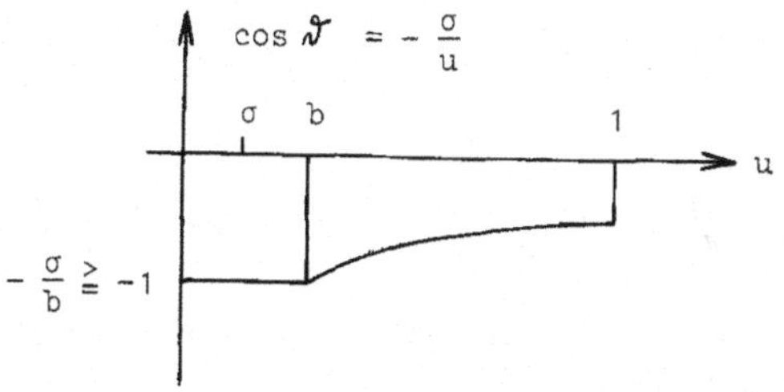

Figure 17.

In the present case we obtain for U:

$$U = -C_1 - \frac{\lambda_1}{2} = 2\sigma - \frac{\lambda_1}{2} = -2\int_b^1 \cos\vartheta \, du = 2\int_b^1 \frac{\sigma}{u}\, du = -2\sigma \log b \geqq ($$

$\Rightarrow$

$$0 \leqq \sigma = -\frac{C_1}{2} = \frac{\lambda_1}{4(1 + \log b)}; \quad \lambda_1 \geqq 0;$$

$$U = -\frac{\lambda_1 \log b}{2(1 + \log b)} \geqq 0.$$

Hence, again, (6) and (8) hold simultaneously if $\lambda_1 \geqq 0$ and

$$\sigma = \frac{|\lambda_1|}{4(1 + \log b)} \leqq b.$$

For determining V and for finding the upper bound for $|\lambda_2|$ the above considerations remain to hold.

Theorem 1. In $S(b)$ and $b \in (e^{-1},1)$ there holds for $\lambda = \lambda_1 + i\lambda_2 \in C$:

$$(12)\quad \operatorname{Re}\,(a_3 + \lambda a_2) \leqq 1 - b^2 - \frac{\lambda_1^2}{4\rho} + \frac{\lambda_2^2}{4}; \quad \rho = 1 + \frac{1}{\log b} < 0.$$

Equality here can be reached only if for λ the following limitations are true:

$$(13)\quad \begin{cases} |\lambda_1| \leqq 4b(1 + \log b), \\ |\lambda_2| \leqq 4(\sqrt{1 - \sigma^2} - \sigma \,\overline{\text{arc}}\cos \sigma - \sqrt{b^2 - \sigma^2} + \sigma\,\overline{\text{arc}}\cos \frac{\sigma}{b}), \end{cases}$$

where

$$(14)\quad 0 \leqq \sigma = \frac{|\lambda_1|}{4(1 + \log b)} \leqq b.$$

For the generating function ϑ, connected with the extremal 2:2-mapping, there holds $(\lambda_1 \neq 0)$

$$(15)\quad -\frac{\lambda_1}{|\lambda_1|}\cos\vartheta = \frac{\sigma}{u}, \quad b \leqq u \leqq 1.$$

If $\lambda_1 = 0$, $\cos\vartheta \equiv 0$ and the extremal mappings have two straight radial slits along the imaginary axis.

In the special case $b = e^{-1}$ the condition (13) gives $\lambda_1 = 0$ and (14) leaves σ undetermined. Thus, extremal functions of the type 2:2 can be expected in the case $b = e^{-1}$ only for purely imaginary values of λ. From (5) we obtain when inserting in it $b = e^{-1}$, $\lambda_1 = 0$ $(\operatorname{Re} x_0 = -C_1 = U + \frac{\lambda_1}{2})$

$$\begin{cases} \operatorname{Re}\,(a_3 + i\lambda_2 a_2) = 1 - e^{-2} + \frac{\lambda_2^2}{4} - (V + \frac{\lambda_2}{2})^2 - 4\int_b^1 X^2 du, \\ X = \sqrt{u}(\cos\vartheta + \frac{U}{2u}). \end{cases}$$

Thus,

$$\operatorname{Re}\,(a_3 + i\lambda_2 a_2) \leqq 1 - e^{-2} + \frac{\lambda_2^2}{4},$$

where the equality is reached for $\cos\vartheta = -\frac{U}{2u}$, $V = -\frac{\lambda_2}{2}$. The condition

$U = -2 \int_b^1 \cos \vartheta \, du$ is reduced to an identity. Hence, there are two free parameters, λ_2 and U, for which

$$|\lambda_2| \leqq 4(1 - e^{-1}), \quad |U| \leqq 2e^{-1}.$$

Thus we see that in the limit process $b \to e^{-1}$ the extremal domains 2:2 valid in the interval $|\lambda_1| \leqq 4b(1 + \log b)$, are all condensed to the points of the line segment $|\lambda_2| \leqq 4(1 - e^{-1})$. At each point of this line segment there lies a one-parametric family of extremal domains 2:2, having U as a free parameter.

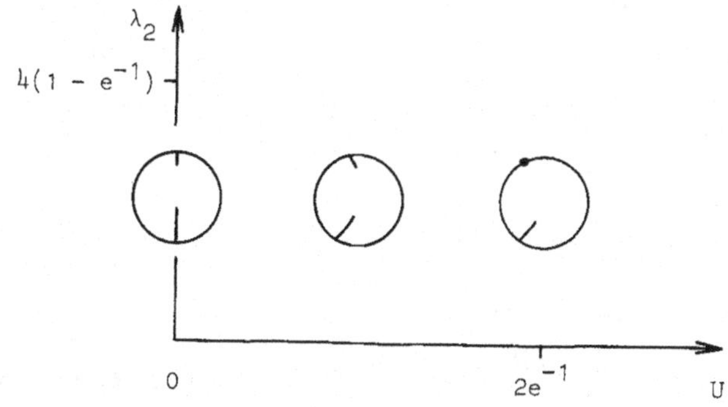

Figure 18.

We can utilize (5) in the special case $\lambda_1 = 0$ to give results behind the limit given by (13) to λ_2: $|\lambda_2| \leqq 4(1 - b)$. Thus, let us first assume that $e^{-1} \leqq b < 1$ and $\lambda_1 = 0$. (5) gives

$$\begin{aligned}
\operatorname{Re}(a_3 + i\lambda_2 a_2) &= 1 - b^2 + \frac{\lambda_2^2}{4} - (V + \frac{\lambda_2}{2})^2 - (1 + \log b)U^2 - 4\int_b^1 (\cos\vartheta + \frac{U}{2u})^2 du \\
&\leqq 1 - b^2 + \frac{\lambda_2^2}{4} - (V + \frac{\lambda_2}{2})^2 \\
&\leqq 1 - b^2 + \frac{\lambda_2^2}{4} - (\pm 2(1 - b) + \frac{\lambda_2}{2})^2 .
\end{aligned}$$

Here + and $\lambda_2 < 0$ as well as - and $\lambda_2 > 0$ belong together, as indicated in Figure 19.

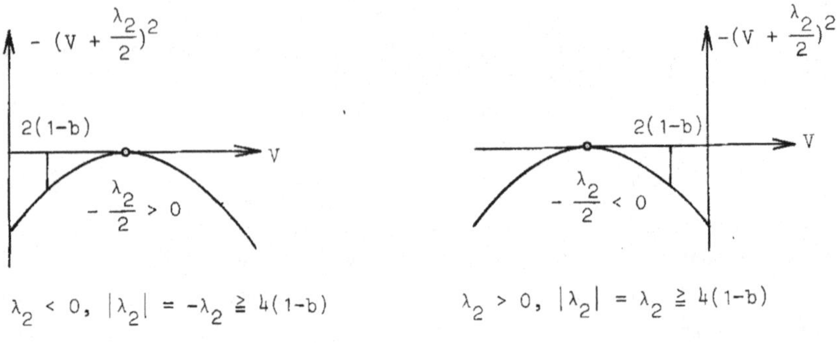

Figure 19.

Theorem 2. In $S(b)$ and $b \in [e^{-1}, 1)$ let $\lambda = \lambda_2 i$ be purely imaginary so that

$$\lambda = i\lambda_2, \quad |\lambda_2| \geqq 4(1-b).$$

For those values of λ

$$(16) \quad \operatorname{Re}(a_3 + i\lambda_2 a_2) \leqq -3 + 8b - 5b^2 + 2(1-b)|\lambda_2|.$$

If $f(z) = b(z + 2(1-b)z^2 + \dots)$ means the left radial-slit mapping, then equality in the above condition is reached for the rotated radial-slit mapping $\tau^{-1}f(\tau z)$, where $\tau = i$ is to be used for $\lambda_2 < 0$ $(a_2 = 2(1-b)i)$ and $\tau = -i$ for $\lambda_2 > 0$ $(a_2 = -2(1-b)i)$.

For $b = e^{-1}$ and $\lambda = i\lambda_2$ so that $|\lambda_2| \leqq 4(1-e^{-1})$

$$(17) \quad \operatorname{Re}(a_3 + i\lambda_2 a_2) \leqq 1 - e^{-2} + \frac{\lambda_2^2}{4}.$$

Equality holds here for a one-parametric family of functions of the type 2:2 for which

$$\cos\vartheta = -\frac{U}{2u}, \quad V = -\frac{\lambda_2}{2}.$$

Here U is a free parameter in the interval $|U| \leqq 2e^{-1}$.

Also for the values $0 < b < e^{-1}$, $\lambda = i\lambda_2$, there exists a lower limit for $|\lambda_2|$ from which upwards the previous inequality (16) remains to hold.

By aid of Schwarz's inequality we obtain the first information of this limit of λ_2. The result will be completed later on in V.3.4.

Because

$$\left(\int_b^1 \cos\vartheta \, du\right)^2 = \left(\int_b^1 \frac{1}{\sqrt{u}} \sqrt{u} \cos\vartheta \, du\right)^2 \leqq \int_b^1 \frac{du}{u} \cdot \int_b^1 u \cos^2\vartheta \, du,$$

we have

$$-4 \int_b^1 u \cos^2\vartheta \, du \leqq \frac{U^2}{\log b}$$

and thus

$$\mathrm{Re}\,(a_3 + i\lambda_2 a_2) = 4\left(\int_b^1 \cos\vartheta \, du\right)^2 - 4\left(\int_b^1 \sin\vartheta \, du\right)^2$$

$$- 2 \int_b^1 u(2\cos^2\vartheta - 1)du + 2\lambda_2 \int_b^1 \sin\vartheta \, du$$

$$\leqq 1 - b^2 + \underbrace{\frac{1 + \log b}{\log b}}_{\rho} U^2 - V^2 - \lambda_2 V = F(U,V).$$

Observe that if $e^{-1} \leqq b < 1$ the number $\rho = 1 + \frac{1}{\log b} \leqq 0$ and we can perform again the previous estimation.

If $0 < b < e^{-1}$ the number $\rho > 0$ and one can check immediately that the free extremal point $U = 0$, $V = -\frac{\lambda_2}{2}$ does not give a local maximum for F. Hence F which is defined in the disc

$$U^2 + V^2 \leqq R^2, \quad R = 2(1 - b),$$

is necessarily maximized on the boundary of this disc, where $U^2 = R^2 - V^2$ and thus

$$F = 1 - b^2 + \rho(R^2 - V^2) - V^2 - \lambda_2 V$$

$$= 1 - b^2 + \rho R^2 + \frac{\lambda_2^2}{4(1 + \rho)} - (1 + \rho)\left[V + \frac{\lambda_2}{2(1 + \rho)}\right]^2.$$

In Figure 20 there is given a schematic presentation indicating how the function $-(1+\rho)[\ \ ]^2$ is maximized in different cases.

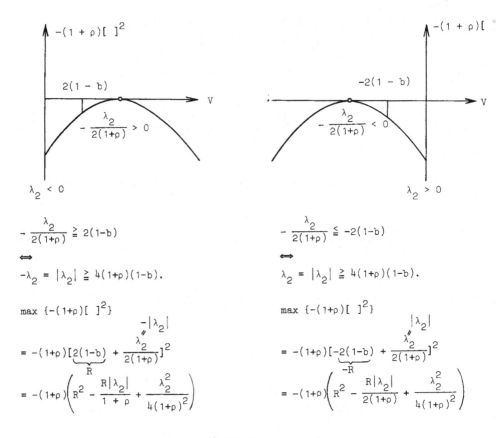

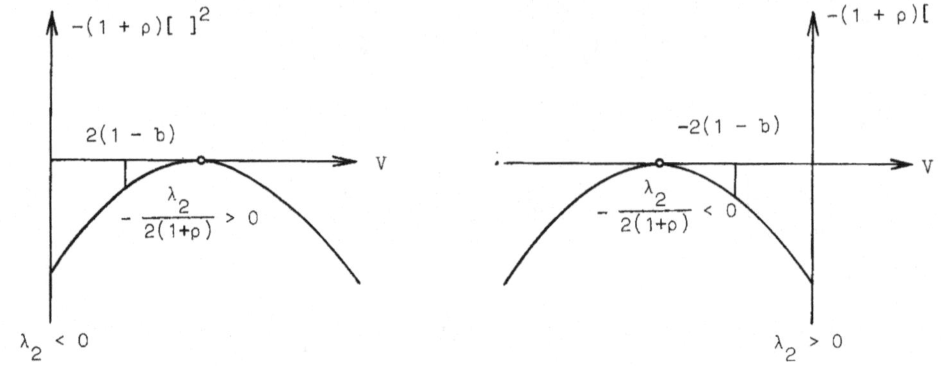

$$-\frac{\lambda_2}{2(1+\rho)} \geqq 2(1-b) \qquad\qquad -\frac{\lambda_2}{2(1+\rho)} \leqq -2(1-b)$$

$$\Longleftrightarrow \qquad\qquad\qquad \Longleftrightarrow$$

$$-\lambda_2 = |\lambda_2| \geqq 4(1+\rho)(1-b). \qquad\qquad \lambda_2 = |\lambda_2| \geqq 4(1+\rho)(1-b).$$

$$\max\{-(1+\rho)[\ \ ]^2\} \qquad\qquad \max\{-(1+\rho)[\ \ ]^2\}$$

$$= -(1+\rho)[\underbrace{2(1-b)}_{R} + \frac{\overset{-|\lambda_2|}{\overset{\|}{\lambda_2}}}{2(1+\rho)}]^2 \qquad\qquad = -(1+\rho)[\underbrace{-2(1-b)}_{-R} + \frac{\overset{|\lambda_2|}{\overset{\|}{\lambda_2}}}{2(1+\rho)}]^2$$

$$= -(1+\rho)\left(R^2 - \frac{R|\lambda_2|}{1+\rho} + \frac{\lambda_2^2}{4(1+\rho)^2}\right) \qquad\qquad = -(1+\rho)\left(R^2 - \frac{R|\lambda_2|}{2(1+\rho)} + \frac{\lambda_2^2}{4(1+\rho)^2}\right)$$

Figure 20.

Independently of the sign of λ_2 we thus have

$$F \leqq 1 - b^2 + \rho R^2 + \frac{\lambda_2^2}{4(1+\rho)} - (1+\rho)R^2 + R|\lambda_2| - \frac{\lambda_2^2}{4(1+\rho)}$$

$$= 1 - b^2 - R^2 + R|\lambda_2| = -3 + 8b - 5b^2 + 2(1-b)|\lambda_2|.$$

Theorem 3. In $S(b)$ and $b \in (0,e^{-1})$ let $\lambda = \lambda_2 i$ be purely imaginary so that

(18) $|\lambda_2| \geqq 4(2 + \frac{1}{\log b})(1 - b)$.

For these values of λ

$$\mathrm{Re}\,(a_3 + i\lambda_2 a_2) \leqq -3 + 8b - 5b^2 + 2(1 - b)|\lambda_2|.$$

Equality holds for the rotated radial-slit function of Theorem 2.

As mentioned above, the limit (18) will be sharpened in V.3.4.

In the previous estimations the choice (4), suggested by the variational condition, was used for the parameter C. In what follows we are going to show that the same results can be obtained also by aid of the optimized inequality from which the coefficient body (a_2, a_3) was determined. In the present case 2:2 we thus start from the estimation (9)/V.1.2:

$$\text{(19)} \qquad \mathrm{Re}\,\delta \leqq 1 - b^2 + \frac{U^2}{\log b}.$$

This gives

$$\mathrm{Re}\,(a_3 + \lambda a_2) \leqq \mathrm{Re}\,(a_2^2 + \lambda a_2) + 1 - b^2 + \frac{U^2}{\log b}$$

$$= 1 - b^2 + \underbrace{(1 + \frac{1}{\log b})}_{\rho}U^2 + \lambda_1 U - V^2 - \lambda_2 V$$

$$= 1 - b^2 + \rho(U + \frac{\lambda_1}{2\rho})^2 - (V + \frac{\lambda_2}{2})^2 - \frac{\lambda_1^2}{4\rho} + \frac{\lambda_2^2}{4}.$$

When looking for the equality cases 2:2 we restrict b so that

$$e^{-1} < b < 1 \Longleftrightarrow \rho = \frac{1 + \log b}{\log b} < 0.$$

– For brevity, the case $b = e^{-1}$ is not treated here. Thus

$$\text{(20)} \qquad \mathrm{Re}\,(a_3 + \lambda a_2) \leqq 1 - b^2 - \frac{\lambda_1^2}{4\rho} + \frac{\lambda_2^2}{4}$$

with the equality if and only if

$$\text{(21)} \qquad U = -\frac{\lambda_1}{2\rho}, \quad V = -\frac{\lambda_2}{2}.$$

Consider the case $\lambda_1 < 0$. In the equality case thus necessarily $U < 0$. From V.1.2, p. 4, we know that in the equality case of (19) and thus (20), we have $x_0 = -C = -C_1 - iC_2$ so optimized that

$$C_1 = -\text{Re } x_0 = \frac{U}{\log b}.$$

Thus, when using this choice for C_1, we have in the equality case (21)

$$(22) \qquad \begin{cases} U = -\dfrac{\lambda_1}{2}\dfrac{\log b}{1 + \log b}, \\[2ex] C_1 = -\dfrac{\lambda_1}{2}\dfrac{1}{1 + \log b}; \end{cases}$$

$$\begin{cases} U + C_1 = -\dfrac{\lambda_1}{2}, \\[2ex] V = -\dfrac{\lambda_2}{2}. \end{cases}$$

This means that $a_2 + C_1 + i\cdot 0 = -\frac{1}{2}\lambda$, i.e.

$$(23) \qquad R \ni -C = a_2 + \frac{1}{2}\lambda.$$

In the case $\lambda_1 > 0$ we have $U > 0$ in the equality case. All the above formulae remain to hold also in this equality case.

From (13)/V.1.2 we know that equality in (19) can be reached only for

$$|U| \leqq 2b|\log b|$$

$\Longleftrightarrow$

$$|\lambda_1| \leqq 4b|\log b|$$

and

$$|V| \leqq 2(\sqrt{1 - \sigma^2} - \sigma \overline{\text{arc}} \cos \sigma - \sqrt{b^2 - \sigma^2} + \sigma \overline{\text{arc}} \cos \frac{\sigma}{b})$$

$\Longleftrightarrow$

$$|\lambda_2| \leqq 4(\sqrt{1 - \sigma^2} - \sigma \overline{\text{arc}} \cos \sigma - \sqrt{b^2 - \sigma^2} + \sigma \overline{\text{arc}} \cos \frac{\sigma}{b})$$

where (the upper sign belongs to $U > 0$, the lower to $U < 0$)

$$\sigma = \mp \frac{U}{2 \log b} = \pm \frac{\lambda_1}{4} \frac{1}{1 + \log b} \in [0,b].$$

Thus, we rederived the essential formulae of Theorem 4/V.3.2 which thus is obtained also by using the optimized inequality.

3. The Extremal Case 1:2

Let us start again from the previous identity (5) which we now write in the form

(24) $$\text{Re}\,(a_3 + \lambda a_2) = 1 - b^2 + \frac{\lambda_2^2}{4} - U^2 - (U + \frac{\lambda_1}{2})^2 \log b - (V + \frac{\lambda_2}{2})^2 - 4\int_b^1 X^2 du;$$

$$X = \sqrt{u}\left(\cos\vartheta + \frac{\text{Re}\,x_0}{2u}\right).$$

First, consider the case $C_1 = -\text{Re}\,x_0 \geqq 0$ in which we denote

$$b \leqq \sigma = \frac{C_1}{2} \leqq 1.$$

In V.1.3 the following sharp estimation was derived:

(25) $$-4\int_b^1 X^2 du \leqq 6\sigma^2 - 4\sigma^2 \log \sigma + 2b^2 - 8b\sigma + 4\sigma^2 \log b$$

with the equality for

(26) $$\cos\vartheta = \begin{cases} 1, & b \leqq u \leqq \sigma, \\ \frac{\sigma}{u}, & \sigma \leqq u \leqq 1. \end{cases}$$

Let us, again, make use of the choice (4), giving

$$\sigma = \frac{C_1}{2} = -\frac{U}{2} - \frac{\lambda_1}{4} \Rightarrow U = -2\sigma - \frac{\lambda_1}{2}.$$

Thus, (24) together with (25) gives

$$\mathrm{Re}\,(a_3 + \lambda a_2) \leqq 1 - b^2 + \frac{\lambda_2^2}{4} - (2\sigma + \frac{\lambda_1}{2})^2 - 4\sigma^2 \log b$$

$$+ 6\sigma^2 - 4\sigma^2 \log \sigma + 2b^2 - 8b\sigma + 4\sigma^2 \log b$$

$$= 1 + b^2 - \frac{\lambda_1^2}{4} + \frac{\lambda_2^2}{4} + 2\sigma^2 - 2\lambda_1\sigma - 8b\sigma - 4\sigma^2 \log \sigma$$

with the equality only for

$$V = -\frac{\lambda_2}{2}.$$

Write this, for brevity:

$$(27)\qquad \begin{cases} \mathrm{Re}\,(a_3 + \lambda a_2) \leqq 1 + b^2 - \dfrac{\lambda_1^2}{4} + \dfrac{\lambda_2^2}{4} + F(\sigma), \\ F(\sigma) = 2\sigma^2 - 2\lambda_1\sigma - 8b\sigma - 4\sigma^2 \log \sigma. \end{cases}$$

We want to fix σ so that $F(\sigma)$ is maximized. Therefore, form the condition

$$(28)\qquad F'(\sigma) = -8(\sigma \log \sigma + b + \frac{\lambda_1}{4}) = 0.$$

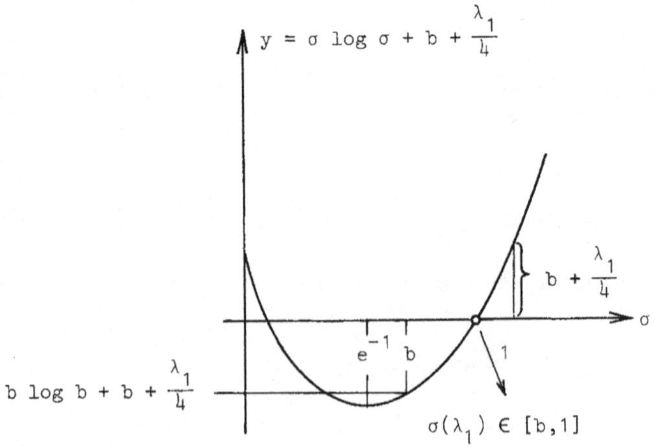

Figure 21.

The equation (28) has a root $\sigma(\lambda_1) \in [b,1]$ if and only if

$$b + \frac{\lambda_1}{4} \geqq 0 \quad \text{and} \quad b \log b + b + \frac{\lambda_1}{4} \leqq 0$$

$\Rightarrow$

(29) $$-4b \leqq \lambda_1 \leqq -4b(1 + \log b).$$

The sign of F' changes at $\sigma(\lambda_1)$ from positive to negative. Therefore

$$\max F = F(\sigma(\lambda_1)).$$

Again, we have to check under which conditions equality in (27) is reachable. As mentioned, in the equality case $V = -\frac{\lambda_2}{2}$. For U we obtain

(30) $$U = -C_1 - \frac{\lambda_1}{2} = -2\sigma - \frac{\lambda_1}{2}$$

$$= -2\int_b^1 \cos\vartheta\, du = -2\int_b^\sigma du - 2\int_\sigma^1 \frac{\sigma}{u}\, du = -2\sigma + 2b + 2\sigma \log \sigma.$$

Because

$$\frac{d}{d\sigma}(-2\sigma + 2b + 2\sigma \log \sigma) = 2 \log \sigma < 0,$$

we have for U the upper bound:

$$U \leqq 2b \log b < 0 \Rightarrow |U| \geqq 2b|\log b|.$$

Because $\lambda_1 U \geqq 0$ in the maximum case, equality in (27) requires that

$$\lambda_1 \leqq 0.$$

If $e^{-1} \leqq b \leqq 1$, the condition (29) implies this limitation automatically. If $0 < b < e^{-1}$, we replace (29) by the condition

(31) $$-4b \leqq \lambda_1 \leqq 0.$$

From (30) we see that in the extremal case

$$\sigma \log \sigma + b + \frac{\lambda_1}{4} = 0.$$

This is the condition (28) which means that $-4\int_b^1 X^2 du$ and F are maximized simultaneously.

For the extremal ϑ we have the conditions (26) and, therefore

$$(32)\qquad \begin{cases} \sin\vartheta = 0 \quad \text{for} \quad b \leqq u \leqq \sigma, \\ \sin\vartheta = \begin{cases} \sqrt{1 - \dfrac{\sigma^2}{u^2}}, & \sigma \leqq u \leqq c, \\ -\sqrt{1 - \dfrac{\sigma^2}{u^2}}, & c \leqq u \leqq 1. \end{cases} \end{cases}$$

By using this and the condition $V = -\frac{\lambda_2}{2}$, we obtain for λ_2 and for V in the extremal case

$$-\frac{\lambda_2}{2} = V = -2\int_b^1 \sin\vartheta \, du = -2\int_\sigma^c \sqrt{1 - \frac{\sigma^2}{u^2}}\, du + 2\int_c^1 \sqrt{1 - \frac{\sigma^2}{u^2}}\, du.$$

This condition can be met with provided that for λ_2 holds

$$(33)\qquad \left|\frac{\lambda_2}{4}\right| \leqq \int_\sigma^1 \sqrt{1 - \frac{\sigma^2}{u^2}}\, du = \sqrt{1 - \sigma^2} - \sigma \,\overline{\text{arc}} \cos \sigma.$$

For the maximal F we obtain

$$F(\sigma(\lambda_1)) = 2\sigma^2 - 2\lambda_1\sigma - 8b\sigma - 4\sigma^2 \log \sigma = 2\sigma^2 - \lambda_1\sigma - 4b\sigma$$

which implies, finally,

$$(34)\qquad \operatorname{Re}(a_3 + \lambda a_2) \leqq 1 + b^2 - \frac{\lambda_1^2}{4} + \frac{\lambda_2^2}{4} + 2\sigma^2 - \lambda_1\sigma - 4b\sigma.$$

Next, let us take $C_1 = -\operatorname{Re} x_0 \leqq 0$ so that

$$b \leqq \sigma = -\frac{C_1}{2} \leqq 1.$$

Now we have, according to V.1.3, for $-4\int_b^1 X^2 du$ the previous upper bound (25), with the equality condition

$$\cos \vartheta = \begin{cases} -1, & b \leqq u \leqq \sigma, \\ -\frac{\sigma}{u}, & \sigma \leqq u \leqq 1. \end{cases}$$

Further

$$\sigma = -\frac{C_1}{2} = \frac{U}{2} + \frac{\lambda_1}{4} \Rightarrow U = 2\sigma - \frac{\lambda_1}{2}$$

and (24) gives by aid of (25)

$$\mathrm{Re}\,(a_3 + \lambda a_2) \leqq 1 - b^2 + \frac{\lambda_2^2}{4} - (2\sigma - \frac{\lambda_1}{2})^2 - 4\sigma^2 \log b$$

$$+ 6\sigma^2 - 4\sigma^2 \log \sigma + 2b^2 - 8b\sigma + 4\sigma^2 \log b$$

$$= 1 + b^2 - \frac{\lambda_1^2}{4} + \frac{\lambda_2^2}{4} + \underbrace{2\sigma^2 + 2\lambda_1\sigma - 8b\sigma - 4\sigma^2 \log \sigma}_{F(\sigma)}$$

with the equality $V = -\frac{\lambda_2}{2}$, as before. Thus we see, that in comparison to the former upper bound (27) the only difference is the sign of λ_1. This means that for the maximizing σ

$$F'(\sigma) = -8(\sigma \log \sigma + b - \frac{\lambda_1}{4}) = 0$$

and this equation has a root $\sigma(\lambda_1) \in [b,1]$ if and only if

$$4b(1 + \log b) \leqq \lambda_1 \leqq 4b.$$

$$U = 2\sigma - \frac{\lambda_1}{2} = 2\int_b^\sigma du + 2\int_\sigma^1 \frac{\sigma}{u}\, du = 2\sigma - 2b - 2\sigma \log \sigma \geqq -2b \log b > 0$$

$$\Rightarrow U \geqq 2b|\log b|.$$

The maximum of $\operatorname{Re}(a_3 + \lambda a_2)$ can be reached only if $\lambda_1 \geqq 0$. Again, the above bounds for λ_1 are automatically correct if $e^{-1} \leqq b < 1$ but if $0 < b < e^{-1}$ we must require

$$0 \leqq \lambda_1 \leqq 4b.$$

Clearly, the formulae (32)-(34) preserve their meaning also in the present case.

<u>Theorem 4</u>. In $S(b)$ and $b \in [e^{-1},1)$ the estimation of $\operatorname{Re}(a_3 + \lambda a_2)$ can be extended for the following values of $\lambda = \lambda_1 + i\lambda_2$:

$$(35) \quad \begin{cases} 4b(1 + \log b) \leqq |\lambda_1| \leqq 4b, \\ |\lambda_2| \leqq 4(\sqrt{1 - \sigma^2} - \sigma \, \overline{\operatorname{arc}} \cos \sigma), \end{cases}$$

where $\sigma \in [e^{-1},1]$ is the root of the equation

$$(36) \quad \sigma \log \sigma + b - \frac{|\lambda_1|}{4} = 0.$$

For these values of λ there holds

$$(37) \quad \operatorname{Re}(a_3 + \lambda a_2) \leqq 1 + b^2 - \frac{\lambda_1^2}{4} + \frac{\lambda_2^2}{4} + 2\sigma^2 + |\lambda_1|\sigma - 4b\sigma.$$

The function ϑ connected with the extremal 1:2-mapping satisfies $(\lambda_1 \neq 0)$ the conditions

$$(38) \quad -\frac{\lambda_1}{|\lambda_1|} \cos \vartheta = \begin{cases} 1, & b \leqq u \leqq \sigma, \\ \dfrac{\sigma}{u}, & \sigma \leqq u \leqq 1. \end{cases}$$

If $b \in (0,e^{-1})$ the above results remain to hold, when λ_1 in (35) is limited so that

$$(39) \quad |\lambda_1| \leqq 4b.$$

If $\lambda_1 = 0$ in this case $b \in (0,e^{-1})$, the extremal function ϑ is further determined by (38) when replacing in it $\lambda_1/|\lambda_1|$ by its right or left limit value $+1$ or -1. Thus, for $\lambda_1 = 0$ we have two extremal domains for each $\lambda = i\lambda_2$ satisfying (35).

$e^{-1} < b = 0.5$

$$(13)-(14)\begin{cases} |\lambda_1| \leqq 4b(1+\log b) = 0.613'706, \\ \sigma = \dfrac{|\lambda_1|}{4(1+\log b)} \leqq b, \\ |\lambda_2^{(1)}| = 4(\sqrt{1-\sigma^2} - \sigma\,\overline{\text{arc}}\cos\sigma - \sqrt{b^2-\sigma^2} + \sigma\,\overline{\text{arc}}\cos\frac{\sigma}{b}). \end{cases}$$

$$(35)-(36)\begin{cases} |\lambda_1| \leqq 4b = 2, \\ \sigma\log\sigma + b - \dfrac{|\lambda_1|}{4} = 0; \quad \sigma \in [e^{-1},1], \\ |\lambda_2^{(2)}| = 4(\sqrt{1-\sigma^2} - \sigma\,\overline{\text{arc}}\cos\sigma). \end{cases}$$

$\lvert\lambda_1\rvert$	σ	$\lvert\lambda_2^{(1)}\rvert$; $\lvert\lambda_2^{(2)}\rvert$	
0	0	2	$\lvert\lambda_2^{(1)}\rvert$
0.1	0.081'472	1.986'673	
0.2	0.162'945	1.946'045	
0.3	0.244'417	1.875'985	
0.4	0.325'889	1.772'101	
0.5	0.407'361	1.625'381	
0.6	0.488'834	1.410'467	
0.613'706	0.5	1.369'707	
0.7	0.559'281	1.129'620	$\lvert\lambda_2^{(2)}\rvert$
0.8	0.612'993	0.926'723	
0.9	0.658'698	0.765'553	
1	0.699'491	0.631'076	
1.1	0.736'868	0.516'030	
1.2	0.771'691	0.416'282	
1.3	0.804'508	0.329'261	
1.4	0.835'695	0.253'285	
1.5	0.865'523	0.187'254	
1.6	0.894'194	0.130'491	
1.7	0.921'865	0.082'693	
1.8	0.948'659	0.043'985	
1.9	0.974'677	0.015'217	
2	1	0	

Table 6.

$$b = 0.25 < e^{-1}$$

$$(35)\text{-}(36)\begin{cases} |\lambda_1| \leqq 4b = 1, \\ \sigma \log \sigma + b - \dfrac{|\lambda_1|}{4} = 0; \quad \sigma \in [e^{-1},1], \\ |\lambda_2^{(2)}| = 4(\sqrt{1-\sigma^2} - \sigma\,\overline{\text{arc}}\cos\sigma). \end{cases}$$

$\lvert\lambda_1\rvert$	σ	$\lvert\lambda_2^{(2)}\rvert$
0	0.699˙491	0.631˙076
0.1	0.736˙868	0.516˙080
0.2	0.771˙691	0.416˙282
0.3	0.804˙508	0.329˙261
0.4	0.835˙695	0.253˙285
0.5	0.865˙523	0.187˙254
0.6	0.894˙194	0.130˙491
0.7	0.921˙865	0.082˙693
0.8	0.948˙659	0.043˙985
0.9	0.974˙677	0.015˙217
1	1	0

Table 7.

In Tables 6 and 7 there are numerical values giving boundary points for the domains, defined by the formulae (13)-(14) and (35)-(36). In these domains, which will be illustrated at the end of V.3.4 (Figures 24 and 25) the extremal functions are of the type 2:2 and 1:2.

As in previous Section V.3.2, we want to show also here that the optimized inequality derived for determining the corresponding part of (a_2,a_3), can be utilized in the present case, too. Thus, we start from the estimation (25)/V.1.3, p. 17:

$$\text{Re}\,\delta \leqq 1 - b^2 - 2|U|\sigma + 2(\sigma - b)^2$$

$\Rightarrow$

(40) $$\text{Re}\,(a_3 + \lambda a_2) \leqq \text{Re}\,(a_2^2 + \lambda a_2) + 1 - b^2 - 2|U|\sigma + 2(\sigma - b)^2.$$

Here $\sigma = \sigma(U) \in [b,1]$ is the root of the equation

(41) $$\sigma \log \sigma - \sigma + b + \frac{|U|}{2} = 0, \quad |U| \geqq 2b|\log b|.$$

Consider the case $\lambda_1 \leqq 0$; $U \leqq 0$, for which thus

$$U = 2(\sigma \log \sigma - \sigma + b).$$

This defines an inverse mapping $\sigma = \sigma(U)$:

$$\sigma\colon [-2(1-b), -2b|\log b|] \to [b,1].$$

Because of this bijective connection we can express the right side of (40) in V and σ

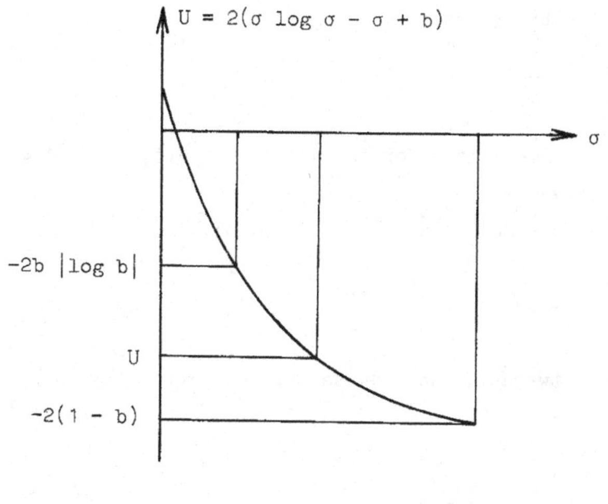

Figure 22.

(42) $$\operatorname{Re}(a_3 + \lambda a_2) \leqq U^2 - V^2 + \lambda_1 U - \lambda_2 V + 1 - b^2 + 2U\sigma + 2(\sigma - b)^2$$

$$= 1 - b^2 + \frac{\lambda_2^2}{4} - (V + \frac{\lambda_2}{2})^2 + 4(\sigma \log \sigma - \sigma + b)^2$$

$$+ 2\lambda_1(\sigma \log \sigma - \sigma + b) + 4\sigma(\sigma \log \sigma - \sigma + b) + 2(\sigma - b)^2.$$

The maximizing value for V is

$$V = -\frac{\lambda_2}{2}$$

and for the maximizing σ we obtain:

$$(43)\quad \frac{\partial}{\partial\sigma} = 8(\sigma\log\sigma-\sigma+b)\log\sigma + 2\lambda_1\log\sigma + \underbrace{4(\sigma\log\sigma-\sigma+b) + 4\sigma\log\sigma + 4(\sigma-b)}_{8\sigma\log\sigma}$$

$$= 8\log\sigma\,(\sigma\log\sigma + b + \frac{\lambda_1}{4}) = 0.$$

This equation is close to that in (28). Thus, we know that it has a root $\sigma = \sigma(\lambda_1) \in [b,1]$ provided that

$$\begin{cases} -4b \leqq \lambda_1 \leqq -4b(1+\log b), & e^{-1} \leqq b < 1; \\ -4b \leqq \lambda_1 \leqq 0, & 0 < b < e^{-1}. \end{cases}$$

The right side of (42) is maximized for the value $\sigma = \sigma(\lambda_1)$, as is seen from the sign of the derivative.

For $U \leqq 0$ we denote according to V.1.3, p. 13,

$$-\mathrm{Re}\, x_0 = C_1 = 2\sigma.$$

The above connections between U and σ as well as between σ and λ_1 give in the maximum case

$$-\frac{\lambda_1}{4} = \sigma\log\sigma + b = \frac{U}{2} + \sigma$$

$\Rightarrow$

$$C_1 = -U - \frac{\lambda_1}{2}.$$

Therefore, in this case we have, as before

$$\begin{cases} U + C_1 = -\frac{\lambda_1}{2}, \\ V = -\frac{\lambda_2}{2}; \end{cases}$$

(44) $\quad R \ni -C = a_2 + \frac{1}{2}\lambda.$

By determining the maximum value of the right side of (42) we obtain the inequality sought:

$$\mathrm{Re}\,(a_3 + \lambda a_2) \leqq 1 - b^2 + \frac{\lambda_2^2}{4} + 4(-\sigma - \frac{\lambda_1}{4})^2 + (2\lambda_1 + 4\sigma)(-\sigma - \frac{\lambda_1}{4}) + 2(\sigma - b)^2$$

$$= 1 + b^2 - \frac{\lambda_1^2}{4} + \frac{\lambda_2^2}{4} + 2\sigma^2 - \lambda_1\sigma - 4b\sigma.$$

If $\lambda_1 \geqq 0$; $U \geqq 0$, we may proceed as before when using the optimized inequality of V.1.3. We have now, in the maximum case, correspondingly:

$$\begin{cases} \mathrm{Re}\; x_0 = -C_1 = 2\sigma, \\ \sigma \log \sigma - \sigma + b + \frac{U}{2} = 0, \\ 4b(1 + \log b) \leqq \lambda_1 \leqq 4b \quad \text{or} \quad 0 \leqq \lambda_1 \leqq 4b, \\ V = -\frac{\lambda_2}{2}, \\ \sigma \log \sigma + b - \frac{\lambda_1}{4} = 0. \end{cases}$$

The only difference in the upper bound is the change of sign of λ_1. Thus, we have generally

(45) $$\mathrm{Re}\,(a_3 + \lambda a_2) \leqq 1 + b^2 - \frac{\lambda_1^2}{4} + \frac{\lambda_2^2}{4} + 2\sigma^2 + |\lambda_1|\sigma - 4b\sigma,$$

which is the previous upper bound (37).

From (28)/V.1.3, p. 17, we see that in the equality case of this condition (45) there holds for λ_2

$$|\lambda_2| \leqq 4(\sqrt{1 - \sigma^2} - \sigma\,\overline{\operatorname{arc}\cos}\,\sigma).$$

This is the latter condition (35). The former ones, either in the form (35) or (39), were also confirmed by the above formulae. Thus, Theorem 4 is also reconfirmed by starting from the optimized inequality, suitable for the equality case 1:2.

4. The Extremal Case 1:1

Finally, we have to consider the combination $\mathrm{Re}\,(a_3 + \lambda a_2)$ for those values of λ that lie in the complement of the range L of λ defined by (13)-(14) and (35)-(36) (Figure 23). We are going to make use of the same method that was succesfully applied at the end of the two previous sections, i.e., we will utilize the inequality (46) of Theorem 5/V.1.4, p. 31. For brevity, let us apply it in the form (54) of Theorem 8/V.1.4, p. 43. Thus, we start from the condition

$$(46)\quad \mathrm{Re}\,\delta \leqq 1 - b^2 + C_1U - C_2V + C_1C_2(\tan\alpha - \tan\omega) - \frac{C_2^2}{2}(\sin^{-2}\alpha - \sin^{-2}\omega).$$

Here

$$(47)\quad \begin{cases} C_1 = 2\,\dfrac{\sin\alpha - b\sin\omega}{\sin(\alpha-\omega)}\cos\alpha\cos\omega, \\[2ex] C_2 = 2\,\dfrac{\cos\alpha - b\cos\omega}{\sin(\alpha-\omega)}\sin\alpha\sin\omega. \end{cases}$$

The numbers α and ω are determined for the coefficient $a_2 = U + iV$, provided that a_2 lies in proper parts of the disc $|a_2| \leqq 2(1-b)$, by the equations

$$(48)\quad \begin{cases} U = C_1\log\dfrac{\cos\alpha}{\cos\omega} + C_2\,(\cot\alpha - \cot\omega + \alpha - \omega), \\[2ex] V = C_1\,(\tan\alpha - \tan\omega - \alpha + \omega) + C_2\log\dfrac{\sin\alpha}{\sin\omega}. \end{cases}$$

The conditions (46)-(48) give for our functional:

$$(49)\quad \mathrm{Re}\,(a_3 + \lambda a_2) \leqq U^2 - V^2 + \lambda_1U - \lambda_2V + 1 - b^2 + C_1U - C_2V$$
$$+ C_1C_2\,(\tan\alpha - \tan\omega) - \frac{C_2^2}{2}(\sin^{-2}\alpha - \sin^{-2}\omega) = F(\alpha,\omega)$$

which is to be maximized in the case that $(U,V) \in \overline{E}$ and, correspondingly, $(\alpha,\omega) \in \overline{T}$. - We assume here that the connection $(\alpha,\omega) \leftrightarrow (U,V)$ between $\overline{T}$ and $\overline{E}$ is bijective (Theorem 7/V.1.4, p. 40). Here T and E are domains defined in Figure 8, p. 46. According to Theorem 6/V.1.4, p. 39, the condition (49) is extended to the boundaries ∂E and ∂T.

Because $F(\alpha,\omega)$ is continuous in the closed set $\overline{T}$, it reaches maximum there. In order to determine the type of the extremal function f we have to consider separately boundary points and inner points of T.

Let us normalize λ so that

$$(50) \qquad \lambda_1 \leqq 0, \quad \lambda_2 \leqq 0.$$

According to V.3.1 we know that in the maximum case

$$(51) \qquad U \leqq 0, \quad V \geqq 0.$$

This means that we are going to consider the triangles T and E which belong to the case 1° in Figure 8, p. 46.

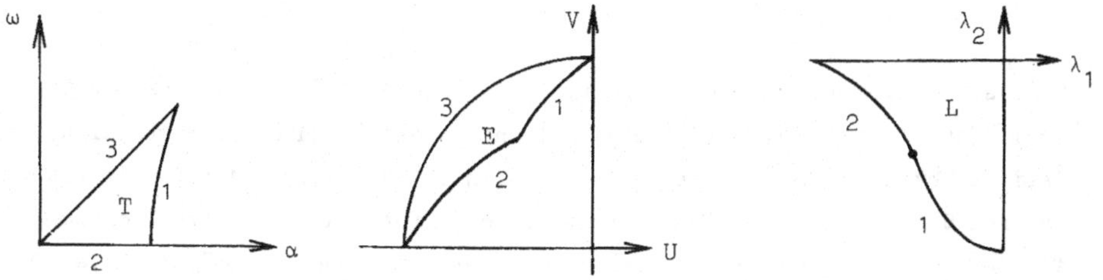

Case 1°

Figure 23.

On the parts 1 and 2 of ∂T we are dealing with the limit cases 2:2 and 1:2 as pointed out in Theorem 6/V.1.4. There we saw also that on the line segment $3 \subset \partial T$ the condition (49) is reduced to an equality with the equality functions of the radial-slit type. We know also that the maximizing ϑ satisfies the variational condition (3)/V.3.1, p. 61. Thus, we can ask which radial-slit mappings $\vartheta = \vartheta(u) =$ constant are defined by this condition (3). For ϑ' it gives

$$(C_1 \cos \vartheta - C_2 \sin \vartheta - 2u \cos 2\vartheta)\, \vartheta' - \sin 2\vartheta = 0.$$

Hence, $\vartheta'(u) \equiv 0$ implies that necessarily $\sin 2\vartheta = 0$, i.e. the only possible radial-slit mappings are those, in which

$$\sin\vartheta \equiv 0,\ C_2 = 0;\ V = \lambda_2 = 0,$$
$$\cos\vartheta \equiv 0,\ C_1 = 0;\ U = \lambda_1 = 0.$$

This means that we are led to the end points of the line segment 3. - The case $\sin\vartheta \equiv 0$ is the radial-slit case we met before in connection of real λ.

We still have to consider those cases in which $\operatorname{Re}(a_3 + \lambda a_2)$ is maximized in E, i.e. for functions of the type 1:1. We know that such extremal points exist, because in L and on its boundary the types 2:2 and 1:2 hold and $\operatorname{Re}(a_3 + \lambda a_2)$ as a continuous functional has maximum also when $\lambda \in -L$. The free extremal point in T satisfies the necessary conditions

(52) $$\frac{\partial F}{\partial \alpha} = \frac{\partial F}{\partial \omega} = 0.$$

This system has a solution of the type 1:1 when $\lambda \in -L$. The explicit form for the conditions (52) can be found, without tedious direct calculations, by using the variational condition (3)/V.3.1, p. 61 together with Theorem 5/V.1.4, p. 31. The extremal inner point (α,ω) we are looking for defines, according to Theorem 5, a generating function ϑ which is necessarily a solution of the variational condition (3) (observe, that the equations (39)/V.1.4, p. 29 imply the variational condition (3)/V.3.1, p. 61). The last two equations of (3) determine connection to the corresponding point $\lambda \in -L$:

$$C_1 = -\left(U + \frac{\lambda_1}{2}\right),\quad C_2 = -\left(V + \frac{\lambda_2}{2}\right).$$

By eliminating U and V from these and (48) we obtain the necessary conditions (54) sought.

<u>Theorem 5</u>. The functions f maximizing $\operatorname{Re}(a_3 + \lambda a_2)$ in S(b) for $\lambda \in -L$ are of the type 1:1. If $a_2 = U + iV$ belongs to the maximizing function f, the signs of U and V are connected with those of λ_1 and λ_2 so that

(53) $$\lambda_1 U \geqq 0,\quad \lambda_2 V \leqq 0.$$

To $a_2 \in E$ there belongs, according to Figure 8, p. 46, the triangle

T, in which the corresponding solution (α,ω) of the following system of equations lies:

$$(54)\quad \begin{cases} (1 + \log \frac{\cos \alpha}{\cos \omega})C_1 + (\cot \alpha - \cot \omega + \alpha - \omega)C_2 + \frac{\lambda_1}{2} = 0, \\ (\tan \alpha - \tan \omega - \alpha + \omega)C_1 + (1 + \log \frac{\sin \alpha}{\sin \omega})C_2 + \frac{\lambda_2}{2} = 0; \\ C_1 = 2\, \frac{\sin \alpha - b \sin \omega}{\sin(\alpha - \omega)} \cos \alpha \cos \omega, \\ C_2 = 2\, \frac{\cos \alpha - b \cos \omega}{\sin(\alpha - \omega)} \sin \alpha \sin \omega. \end{cases}$$

This solution determines

$$(55)\quad \max \operatorname{Re}(a_3 + \lambda a_2) = 1 - b^2 - \frac{\lambda_1^2}{4} + \frac{\lambda_2^2}{4} - \frac{1}{2}\lambda_1 C_1 + \frac{1}{2}\lambda_2 C_2$$
$$+ C_1 C_2 (\tan \alpha - \tan \omega) - \frac{1}{2} C_2^2 (\sin^{-2}\alpha - \sin^{-2}\omega).$$

The radial-slit cases deserve to be considered more closely. As was pointed out on p. 86 in them either $\lambda_2 = 0$ or $\lambda_1 = 0$.

The first case is easily solved by taking $\lambda_2 = 0$ in the general expression (1)/V.3.1, p. 60:

$$\operatorname{Re}(a_3 + \lambda_1 a_2) = U^2 - V^2 - 2\int_b^1 u \cos 2\vartheta \, du + \lambda_1 U$$
$$\leqq U^2 - 2\int_b^1 u \cos 2\vartheta \, du + \lambda_1 U,$$

with the equality if and only if $V = 0$. This upper bound is actually our functional in the real class $S_R(b)$ where we have found the solution in V.2.2 and V.2.3. Especially, if $|\lambda_1| \geqq 4b$ the extremal function is thus a radial slit mapping with a horizontal slit.

If $\lambda_1 = 0$ and $e^{-1} \leqq b < 1$ the complete solution is given by Theorem 1/V.3.2, p. 66, and Theorem 2/V.3.2, p. 68. If $\lambda_1 = 0$ and $0 < b < e^{-1}$ Theorem 4/V.3.3, p. 78, takes care of the values

$$(56)\quad \begin{cases} |\lambda_2| \leqq 4(\sqrt{1 - \sigma^2} - \sigma \operatorname{arc} \cos \sigma); \\ \sigma \log \sigma + b = 0. \end{cases}$$

Further, in Theorem 3/V.3.2, p. 70, it was pointed out that the radial-slit mapping with a vertical slit is the extremal function at least if $\lambda_1 = 0$, $0 < b < e^{-1}$, and

$$|\lambda_2| \geqq 4(2 + \frac{1}{\log b})(1 - b).$$

By aid of the previous result of Theorem 5 we can now sharpen this bound of $|\lambda_2|$.

Let us normalize the problem so that λ_2 is positive. We know that in the corresponding maximum case $V \leqq 0$ and thus we are led to case 3° of Figure 8, p. 46. Numerical calculations show that if we let λ_2 grow from the value defined by (56), the solution (α, ω) of (54) tends to the point $(\frac{3\pi}{2}, \frac{3\pi}{2})$. In order to obtain the limit value of λ_2 which still gives a solution for (54) we perform the following change of variables

$$\alpha = \frac{3\pi}{2} - u, \quad \omega = \frac{3\pi}{2} - v; \quad u < v.$$

- Previously u is used in different sense in Löwner's expressions. The equations (54) assume the form

(57)
$$\begin{cases} \frac{1}{2} C_1 = \frac{\cos u - b \cos v}{\sin (u - v)} \sin u \sin v, \\ \\ \frac{1}{2} C_2 = \frac{\sin u - b \sin v}{\sin (u - v)} \cos u \cos v; \end{cases}$$

(58)
$$(1 + \log \frac{\sin u}{\sin v})(\cos u - b \cos v)\sin u \sin v$$
$$+ (\tan u - \tan v - u + v)(\sin u - b \sin v)\cos u \cos v = 0,$$

(59)
$$[\sin (v - u) + (u - v)\sin u \sin v](\cos u - b \cos v)$$
$$+ (1 + \log \frac{\cos u}{\cos v})(\sin u - b \sin v)\cos u \cos v + \frac{\lambda_2}{4} \sin (u - v) = 0$$

The condition (59) defines the Taylor-expression in u and v:

$$(\frac{\lambda_2}{4} + b)u - (\frac{\lambda_2}{4} + 2b - 1)v$$

$$- (\frac{\lambda_2}{24} + \frac{3+b}{6})u^3 + (\frac{\lambda_2}{8} + \frac{b}{2})u^2 v - (\frac{\lambda_2}{8} + \frac{1}{2})uv^2 + (\frac{\lambda_2}{4} + \frac{5b-1}{6})v^3 + \ldots = 0.$$

From this we solve u in v:

$$(60)\qquad \begin{cases} u = kv + \ell v^3 + \ldots; \\ k = 1 - \dfrac{1 - b}{\dfrac{\lambda_2}{4} + b}, \\ \ell = \dfrac{1}{6}(k-1)^3 + \dfrac{1}{2}\,\dfrac{k^3 + (1-b)k + \frac{1}{3}(1-4b)}{\dfrac{\lambda_2}{4} + b}. \end{cases}$$

This connection (60) can now be used in writing (58) in the form of the Taylor expansion in v:

$$mv^2 - nv^4 + \ldots = 0,$$

which further gives the first approximation for the solution v:

$$(61)\qquad \begin{cases} v = \sqrt{\dfrac{m}{n}}, \\ m = (1-b)k(1 + \log k), \\ n = \dfrac{1}{3}(1-k^3)(k-b) + \dfrac{k}{2}(1+\log k)(k^2 - b) - (1-b)[\ ]_o, \\ [\ ]_o = (\ell - \dfrac{k^3}{6})(2 + \log k) - \dfrac{k}{6}\log k. \end{cases}$$

In Table 8 there are some (u,v)-solutions determining approximately a straight line through the origin with the slope $\frac{v}{u} = 2.718$. The numbers u,v above the horizontal line in the Table are determined from the corresponding (α,ω)-solutions of (54) while the (u,v)-numbers below the line are those given by the first approximation (60)-(61).

The signs of m and n show that the (u,v)-solution exists as far as $m \leqq 0 \Leftrightarrow k \leqq e^{-1} \Rightarrow (1-k)^{-1} \leqq (1-e^{-1})^{-1}$. Because, according to (60),

$$\frac{\lambda_2}{4} + b = \frac{1-b}{1-k} \leqq \frac{1-b}{1-e^{-1}}$$

we obtain for λ_2:

$$0 \leqq \lambda_2 \leqq -4b + 4\,\frac{1-b}{1-e^{-1}} = 4\,\frac{e - 2eb + b}{e - 1}.$$

$b = 0.25, \quad \lambda_1 = 0$

λ_2	u	v
3.71	0.081˙110	0.220˙560
3.745	0.013˙054	0.035˙484
3.745˙5	0.008˙873	0.024˙118
3.745˙8	0.004˙552	0.012˙375
3.745˙85	0.003˙831	0.010˙415
3.745˙90	0.002˙349	0.006˙386
3.745˙93	0.000˙149	0.000˙404
3.745˙930˙1	0.000˙061˙440	0.000˙167˙012
3.745˙930˙120˙6	0.000˙000˙711	0.000˙001˙934

Table 8.

Let us still collect the results concerning the coordinate axes of the $\lambda_1\lambda_2$-plane.

Theorem 6. Consider the special cases of the functional $\operatorname{Re}(a_3 + \lambda a_2)$ where $\lambda = \lambda_1 + i\lambda_2$ is either purely imaginary or real.

If $e^{-1} \leqq b < 1$ the extremal domains on the λ_2-axis consist of two-radial-slit mappings with vertical slits, according to Theorems 1 and 2/V.3.2.

If $0 < b < e^{-1}$ the extremal domains on the λ_2-axis are of the type 1:2 (two solutions, given by the rotation $\tau = -1$ from each other) for the values

$$(62) \quad \begin{cases} |\lambda_2| \leqq 4(\sqrt{1 - \sigma^2} - \sigma \,\overline{\operatorname{arc}}\cos\sigma), \\ \sigma \log \sigma + b = 0. \end{cases}$$

From this onwards, for the values

$$(63) \quad |\lambda_2| < 4\,\frac{e - 2eb + b}{e - 1},$$

the extremal domains are of the type 1:1 having a curved slit determined by the solution of (54). For the values

(64) $|\lambda_2| \geqq 4 \frac{e - 2eb + b}{e - 1}$

the extremal domain is of radial-slit type with a vertical slit.

For $b \in (0,1)$ and $|\lambda_1| \geqq 4b$ the extremal domains on the λ_1-axis are radial-slit domains having a horizontal slit.

In Figures 24 and 25 the types of the extremal domains are presented schematically in the cases $b = 0.5$ and $b = 0.25$.

5. Special Problems

For $b = 0$ and $\lambda = i$ we obtain the Schober-problem concerning maximizing the functional $\text{Re}\,(a_3 + ia_2)$ in S [3]. Actually, a discussion with G. Schober in 1978 give reason to consider his functional and its generalizations.

The system (54) has in the case $b = 0$, $\lambda = i$ the following solutions which are classified according to the cases in Figure 8, p. 46.

3° $\pi < \omega < \alpha < \frac{3\pi}{2}$; $U > 0$, $V < 0$.

$$\begin{cases} \alpha = 3.670\cdot106\cdot186, \\ \omega = 3.207\cdot936\cdot738; \end{cases}$$

$$\begin{cases} U = 1.948\cdot867\cdot088, \\ V = -0.370\cdot514\cdot172. \end{cases}$$

2° $-\frac{\pi}{2} < \alpha' < \omega' < 0$; $U' < 0$, $V' < 0$.

$$\begin{cases} \alpha' = \pi - \alpha = -0.528\cdot513\cdot532, \\ \omega' = \pi - \omega = -0.066\cdot344\cdot080; \end{cases}$$

$U' = -U$, $V' = V$.

Both these solutions give

$$\max \text{Re}\,(a_3 + ia_2) = 3.190\cdot298\cdot109.$$

We observe that in Schober's problem, where λ lies in the line segment

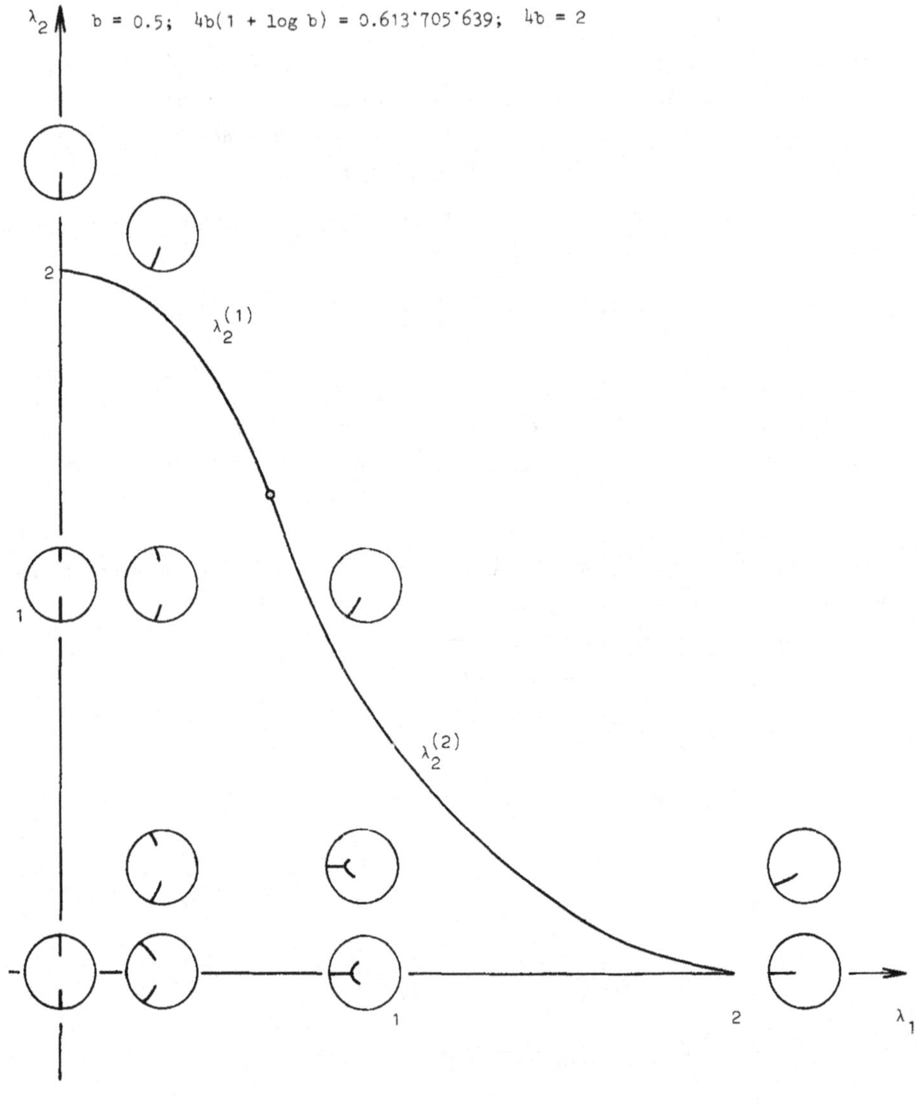

λ_2
b = 0.5; 4b(1 + log b) = 0.613˙705˙639; 4b = 2
2
$\lambda_2^{(1)}$
1
$\lambda_2^{(2)}$
1
2
λ_1

Figure 24.

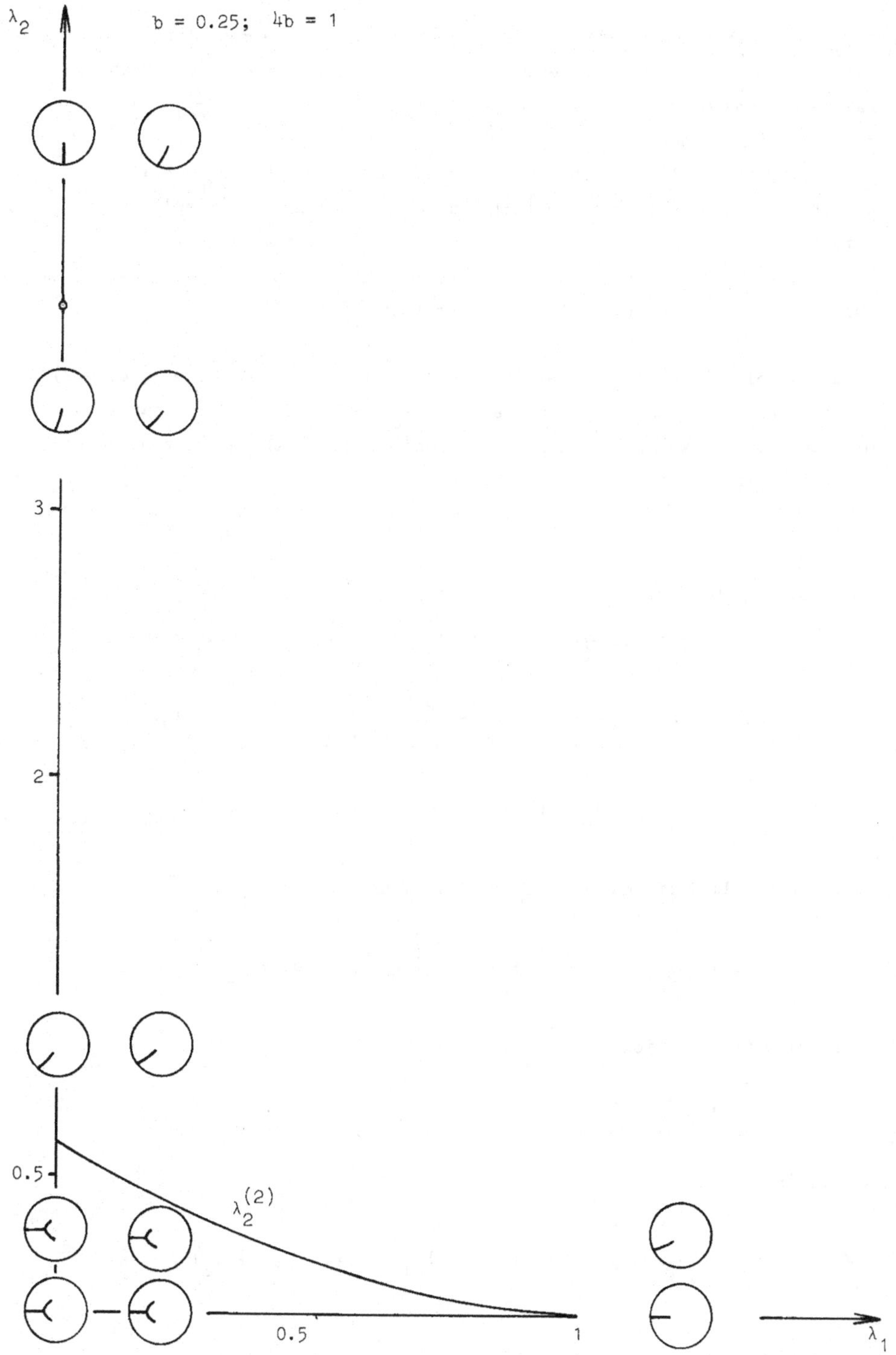

Figure 25.

$[0,\frac{4e}{e-1}]$ of the λ_2-axis, the extremal function is of the most difficult type 1:1 with a curved slit. However, as we noticed earlier, showing that the radial-slit-mapping maximizes $\mathrm{Re}\,(a_3 + i\lambda_2 a_2)$ when $|\lambda_2| \geqq \frac{4e}{e-1}$ is by no means trivial either.

Also if λ lies in the real axis estimation of $\mathrm{Re}\,(a_3 + \lambda_1 a_2)$ is not trivial in $S(b)$, because $|a_3|$ and $|a_2|$ are not maximized simultaneously there. Only when $b = 0$, we can estimate directly:

$$(65) \qquad f \in S\colon \mathrm{Re}\,(a_3 + \lambda a_2) \leqq |a_3 + \lambda a_2| \leqq 3 + 2|\lambda| .$$

Equality holds here for the left Koebe-function if $\lambda > 0$ $(a_3 = 3,\ a_2 = 2)$ and for the right one if $\lambda < 0$ $(a_3 = 3,\ a_2 = -2)$. It appears that in the minimum problem, which we are going to consider next, there are some cases where the direct estimation is possible.

6. The Minimum

Consider first the special case $\lambda_1 = 0;\ \lambda = i\lambda_2$ for which we have

$$\mathrm{Re}\,(a_3 + i\lambda_2 a_2) = U^2 - V^2 - (1 - b^2) + 4\int_b^1 u \sin^2\vartheta\, du - \lambda_2 V$$

$$\geqq -[1 - b^2 + V^2 - 4\int_b^1 u \sin^2\vartheta\, du + \lambda_2 V],$$

with the equality if and only if $U = 0$. Thus

$$-\mathrm{Re}\,(a_3 + i\lambda_2 a_2) \leqq 1 - b^2 + V^2 - 4\int_b^1 u \sin^2\vartheta\, du + \lambda_2 V.$$

When using the notation

$$\vartheta' = \frac{\pi}{2} - \vartheta$$

we obtain from this

$$(66) \quad -\mathrm{Re}\,(a_3 + i\lambda_2 a_2) \leqq 1 - b^2 + U'^2 - 4\int_b^1 u \cos^2\vartheta'\, du + \lambda_2 U',$$

$$U' = -2 \int_b^1 \cos\vartheta' \, du.$$

Hence, we are led to our previous problem of maximizing $a_3 + \lambda a_2$, $\lambda \in R$, in $S_R(b)$. The previous upper bounds (8)/V.2.2, p. 55, and (10)/V.2.3, p. 56, give therefore maximum for $-\mathrm{Re}\,(a_3 + i\lambda_2 a_2)$ and thus minimum for $\mathrm{Re}\,(a_3 + i\lambda_2 a_2)$.

As an example take the class S where $b = 0$. There we have from (65):

$$\mathrm{Re}\,(a_3 + i\lambda_2 a_2) \geqq -3 - 2|\lambda_2|$$

where the equality for $\lambda_2 \geqq 0$ is attained by the left Koebe-function rotated by the factor $\tau = i^{-1}$ $(a_3 = -3,\ a_2 = 2i)$. If $\lambda_2 \leqq 0$ the left Koebe-function rotated by $\tau = i$ $(a_3 = -3,\ a_2 = -2i)$ is the equality function.

For the Schober-functional especially there holds

$$\min_S \mathrm{Re}\,(a_3 + ia_2) = -5.$$

Next, consider the general case where λ is complex and write

(67) $$\mathrm{Re}\,(a_3 + \lambda a_2) = 1 - b^2 - 4\int_b^1 u\cos^2\vartheta\,du$$
$$+ 4\Big(\int_b^1 \cos\vartheta\,du\Big)^2 - 4\Big(\int_b^1 \sin\vartheta\,du\Big)^2 - 2\lambda_1\int_b^1 \cos\vartheta\,du + 2\lambda_2\int_b^1 \sin\vartheta\,du;$$

$$-\mathrm{Re}\,(a_3 + \lambda a_2) = 1 - b^2 - 4\int_b^1 u\sin^2\vartheta\,du$$
$$+ 4\Big(\int_b^1 \sin\vartheta\,du\Big)^2 - 4\Big(\int_b^1 \cos\vartheta\,du\Big)^2 - 2\lambda_2\int_b^1 \sin\vartheta\,du + 2\lambda_1\int_b^1 \cos\vartheta\,du.$$

Again, perform the change of notation

$$\vartheta' = \frac{\pi}{2} - \vartheta,$$

which gives

$$(68)\quad -\mathrm{Re}\,(a_3 + \lambda a_2) = 1 - b^2 - 4\int_b^1 u \cos^2 \vartheta'\, du$$

$$+ 4\,(\int_b^1 \cos \vartheta'\, du)^2 - 4\,(\int_b^1 \sin \vartheta'\, du)^2 - 2\lambda_2 \int_b^1 \cos \vartheta'\, du + 2\lambda_1 \int_b^1 \sin \vartheta'\, d$$

When minimizing the functional (67) we can now proceed as follows. First, change the role of λ_1 and λ_2 and then maximize the corresponding functional (68). The result M which is maximum for $-\mathrm{Re}\,(a_3 + \lambda a_2)$ thus gives:

$$\min \mathrm{Re}\,(a_3 + \lambda a_2) = -M.$$

As an example consider again the problem

$$\min_S \mathrm{Re}\,(a_3 + i\lambda_2 a_2).$$

Because $\lambda = 0 + i\lambda_2$ we change it to the form $\lambda_2 + i0$ and obtain

$$\max_S \mathrm{Re}\,(a_3 + \lambda_2 a_2) = 3 + 2|\lambda_2| = M;$$

$$\min_S \mathrm{Re}\,(a_3 + i\lambda_2 a_2) = -M = -3 - 2|\lambda_2|.$$

The following problem, finally, combines estimations in both directions. We ask, for which values of λ equality is reached simultaneously in either condition

$$(69)\qquad -M \leqq \mathrm{Re}\,(a_3 + \lambda a_2) \leqq M.$$

The solution follows by aid of two observations:

$$f(z) = \sum_1^\infty b_\nu z^\nu \in S(b) \Rightarrow \overline{f}(z) = \sum_1^\infty \overline{b}_\nu z^\nu \in S(b),$$

$$f(z) \in S(b) \Rightarrow i^{-1}\overline{f}(iz) = \sum_1^\infty i^{\nu-1}\overline{b}_\nu z^\nu \in S(b).$$

If, as usual, $b_\nu/b = a_\nu$ then $i^{-1}\overline{f}(iz)$ has the following two first a_ν-coefficients: $i\overline{a}_2$ and $-\overline{a}_3$. Therefore, if

$$\max \operatorname{Re} (a_3 + \lambda a_2) = M,$$

the following sequence of inequalities holds:

$$\operatorname{Re} (a_3 + \lambda a_2) \leqq M$$

$\Rightarrow$

$$\operatorname{Re} (-\overline{a}_3 + \lambda i\overline{a}_2) \leqq M$$

$\Rightarrow$

$$\operatorname{Re} (-a_3 - \overline{\lambda} i a_2) \leqq M$$

$\Rightarrow$

$$\operatorname{Re} (a_3 + i\overline{\lambda} a_2) \geqq -M.$$

This shows that equalities are reached in (69) simultaneously, provided th

$$\lambda = \overline{\lambda} i$$

$\Rightarrow$

$$\lambda = r(1 + i), \quad r \in R.$$

If one takes e.g. $r = 1$; $\lambda = 1 + i$ and $b = 0$ one obtains in S

$$M = 5.153'056'667.$$

The corresponding (α,ω)-values giving the maximal M are

$$\begin{cases} \alpha_1 = 3.468'230'953, \\ \omega_1 = 3.220'639'822. \end{cases}$$

This solution belongs to the case 3° of Figure 8, p. 46, where

$$U > 0, \quad V < 0.$$

The numbers

$$\begin{cases} \alpha_2 = \frac{3\pi}{2} - \alpha_1 = 1.244'154'027, \\ \omega_2 = \frac{3\pi}{2} - \omega_1 = 1.491'749'158, \end{cases}$$

give the minimum $-M$. The order of the numbers is explained by the fact that in the minimum problem for $\mathrm{Re}\,(a_3 + \lambda a_2)$ we replace ϑ by the number $\vartheta' = \frac{\pi}{2} - \vartheta$ in order to find the corresponding maximum problem for $-\mathrm{Re}\,(a_3 + \lambda a_2)$. The solution (α,ω) is thus

$$\begin{cases} \alpha = \frac{\pi}{2} - \alpha_2 = \alpha_1 - \pi, \\ \omega = \frac{\pi}{2} - \omega_2 = \omega_1 - \pi. \end{cases}$$

This leads us to case 1° of Figure 8, where

$$U < 0, \quad V > 0.$$

This, again, is explained by the change of the sign of the functional, as indicated above.

Some of the results of V.2 and V.3 are included in those of [5]. Especially for the extremal case 1:1 the method given in [5] is independent of the perfect square estimation used above.

4 § Determination of Extremal Domains

1. Integration of Löwner's Differential Equation to Schiffer's Equation

In all previous sharp estimations of Löwner-expressions the extremal generating function $\kappa = e^{i\vartheta}$ was determined. The corresponding Löwner's differential equation is

$$(1) \qquad u\frac{\partial f}{\partial u} = f\,\frac{1+\kappa f}{1-\kappa f}, \quad b \leqq u \leqq 1.$$

Its solution

$$f(z,u) = u(z + a_2(u)z^2 + \dots), \quad f(z,1) = z,$$

yields the extremal function $f(z,b)$. It appears that integration with respect to u can be replaced by that with respect to z. This procedure, which essentially facilitates calculations, was found by O. Jokinen [6] and will be described here first.

Lemma 1. With f also $(\bar{f})^{-1}$ is a solution of (1).

This is seen by writing (1) in the form

$$u\frac{\partial f^{-1}}{\partial u} = f^{-1}\,\frac{1+\bar{\kappa}f^{-1}}{1-\bar{\kappa}f^{-1}}$$

and by applying conjugation on both sides.

In IV.2.1 of [1] the factorized differential equation (2)/IV.2.1 was utilized in classifying and representing the boundary functions of the coefficient body (a_2,a_3). In the following we are going to integrate Löwner's equation to the form corresponding to (2)/IV.2.1. This explains introduction of the numbers x_i which are related to the 'singularities of the factorized integrated equation.

Lemma 2. Introduce the differentiable functions

$$x_i = x_i(u) \quad (b \leqq u \leqq 1; \;\; i = 1,\dots,4m; \;\; m = 1,\tfrac{3}{2},2,\tfrac{5}{2},\dots)$$

which satisfy

$$(2) \quad x_1 = x_2, \; |x_1(u)| = 1, \;\; x_i(u) \neq 0, \;\; x_i(u) \neq x_1(u) \;\; (i = 3,\dots,4m);$$

$$(3)\qquad u\,\frac{dx_i}{du} = x_i\,\frac{1+\kappa x_i}{1-\kappa x_i},\quad \kappa = \bar{x}_1 \quad (i = 3,\dots,4m);$$

$$(4)\qquad \frac{d}{du}\prod_{i=1}^{4m} x_i = 0.$$

The following identity then holds

$$(5)\qquad \frac{\partial}{\partial z}\,[u^{m-1}(z^{1/2} + x_1 z^{-1/2})\prod_{i=3}^{4m}(z^{1/2} - x_i z^{-1/2})^{1/2}]$$

$$= \frac{\partial}{\partial u}\,[u^m z^{-1}\prod_{i=1}^{4m}(z^{1/2} - x_i z^{-1/2})^{1/2}].$$

Proof. From (4) we see that Πx_i = constant $\Rightarrow$

$$\sum_1^{4m}\frac{x_i'}{x_i} = 0$$

and from this it follows by using (3)

$$2u\,\frac{x_1'}{x_1} = -\sum_{i=3}^{4m} u\,\frac{x_i'}{x_i} = -\sum_{i=3}^{4m}\frac{1+\bar{x}_1 x_i}{1-\bar{x}_1 x_i};$$

$$(6)\qquad 2u\,\frac{x_1'}{x_1} + \sum_{i=3}^{4m}\frac{x_1+x_i}{x_1-x_i} = 0.$$

By aid of (6) we can now derive the identity needed.

$$2m-1 = \underbrace{2m-1}_{\sum_3^{4m}\frac{1}{2}} + \frac{x_1}{z-x_1}\underbrace{\left(2u\,\frac{x_1'}{x_1} + \sum_{i=3}^{4m}\frac{x_1+x_i}{x_1-x_i}\right)}_{0}$$

$$= \frac{2u\,x_1'}{z-x_1} + \sum_{i=3}^{4m}\left(\frac{1}{2} + \frac{x_1}{z-x_1}\cdot\frac{x_1+x_i}{x_1-x_i}\right)$$

$$= \frac{2u\,x_1'}{z-x_1} + \sum_{i=3}^{4m}\frac{1}{2(z-x_i)}\left[\frac{z+x_1}{z-x_1}(z+x_i) + \frac{x_1+x_i}{x_1-x_i}\,2x_i\right.$$

$$= \frac{2u\,x_1'}{z-x_1} + \sum_{i=3}^{4m} \frac{1}{2}\,\frac{z+x_i}{z-x_i}\,\frac{z+x_1}{z-x_1} + \sum_{i=3}^{4m} \underbrace{x_i\,\frac{x_1+x_i}{x_1-x_i}}_{ux_i'}\,\frac{1}{z-x_i}$$

$$= \sum_{i=3}^{4m} \frac{1}{2}\,\frac{z+x_i}{z-x_i}\,\frac{z+x_1}{z-x_1} + \sum_{i=1}^{4m} \frac{ux_i'}{z-x_i}$$

$\Rightarrow$

$$\frac{1}{2} + \sum_{i=3}^{4m} \frac{1}{4}\,\frac{z+x_i}{z-x_i}\,\frac{z+x_1}{z-x_1} = m + u\sum_{i=1}^{4m} \frac{-x_i'}{2(z-x_i)}.$$

Multiply both sides by $u^{m-1}z^{-1}\prod_{i=1}^{4m}(z^{1/2}-x_iz^{-1/2})^{1/2}$:

$$\frac{1}{2}\,z^{-1}u^{m-1}(z^{1/2}-x_1z^{-1/2})\prod_{i=3}^{4m}(z^{1/2}-x_iz^{-1/2})^{1/2}$$

$$+\, u^{m-1}z^{-1}(z^{1/2}+x_1z^{-1/2})\prod_{i=3}^{4m}(z^{1/2}-x_iz^{-1/2})^{1/2}\sum_{j=3}^{4m}\frac{1}{4}\,\frac{z^{1/2}+x_jz^{-1/2}}{z^{1/2}-x_jz^{-1/2}}$$

$$= mu^{m-1}z^{-1}\prod_{j=1}^{4m}(z^{1/2}-x_iz^{-1/2}) + u^mz^{-1}\prod_{i=1}^{4m}(z^{1/2}-x_iz^{-1/2})^{1/2}\sum_{j=1}^{4m}\frac{-x_j'z^{-1/2}}{2(z^{1/2}-x_jz^{-1/2})}$$

$$u^{m-1}\frac{\partial}{\partial z}(z^{1/2}+x_1z^{-1/2})\prod_{i=3}^{4m}(z^{1/2}-x_iz^{-1/2})^{1/2}$$

$$+\, u^{m-1}(z^{1/2}+x_1z^{-1/2})\frac{\partial}{\partial z}\prod_{i=3}^{4m}(z^{1/2}-x_iz^{-1/2})^{1/2}$$

$$= \frac{\partial}{\partial u}(u^m)\cdot z^{-1}\prod_{i=1}^{4m}(z^{1/2}-x_iz^{-1/2})^{1/2} + u^mz^{-1}\frac{\partial}{\partial u}\prod_{i=1}^{4m}(z^{1/2}-x_iz^{-1/2})^{1/2}.$$

This is the identity (5) sought. □

By using Lemma 2 we now prove:

<u>Theorem 1</u>. Let x_i stand for the functions defined in Lemma 2. Let $f(z,u)$ be the solution of Löwner's differential equation (1) and denote

$$(7)\begin{cases} F(z,u) = \displaystyle\int_{x_3}^{f(z,u)} u^m z^{-1} \prod_{i=1}^{4m} (z^{1/2} - x_i z^{-1/2})^{1/2} dz, \\ |x_3(u)| \leq 1. \end{cases}$$

Then there holds

$$(8) \quad \frac{d}{du} F(z,u) = 0$$

which gives

$$(9) \quad F(f,b) = F(z,1).$$

This is Löwner's equation integrated in the form of Schiffer's equation.

Proof. Write (1) in the form $(x_1 = \bar{\kappa})$

$$u(f^{1/2} - x_1 f^{-1/2}) f^{-1} \frac{\partial f}{\partial u} + (f^{1/2} + x_1 f^{-1/2}) = 0.$$

Multiplication by $u^{m-1} \prod_{i=3}^{4m} (f^{1/2} - x_i f^{-1/2})^{1/2}$ gives

$$u^m f^{-1} \prod_{i=1}^{4m} (f^{1/2} - x_i f^{-1/2})^{1/2} \frac{\partial f}{\partial u}$$

$$+ u^{m-1} (f^{1/2} + x_1 f^{-1/2}) \prod_{i=3}^{4m} (f^{1/2} - x_i f^{-1/2})^{1/2} = 0$$

$\Rightarrow$

$$\frac{\partial}{\partial f}\left[\int_{x_3}^{f} u^m z^{-1} \prod_{i=1}^{4m} (z^{1/2} - x_i z^{-1/2})^{1/2} dz\right] \frac{\partial f}{\partial u}$$

$$+ \int_{x_\sigma}^{f} \frac{\partial}{\partial z}\left[u^{m-1}(z^{1/2} + x_1 z^{-1/2}) \prod_{i=3}^{4m} (z^{1/2} - x_i z^{-1/2})^{1/2}\right] dz = 0.$$

Here also the integral $\int_{x_3}^{f} \frac{\partial}{\partial z}[\ ]dz$ converges at the point $z = x_3$. By aid of (5) this is equivalent to

$$(10)\qquad \frac{\partial}{\partial f}\left[\int_{x_3}^{f} u^m z^{-1}\prod_{i=1}^{4m}(z^{1/2} - x_i z^{-1/2})^{1/2}dz\right]\frac{\partial f}{\partial u}$$

$$+ \int_{x_3}^{f} \frac{\partial}{\partial u}\left[u^m z^{-1}\prod_{i=1}^{4m}(z^{1/2} - x_i z^{-1/2})\right]dz = 0.$$

The latter integral allows changing the order of differentiation and integratic which implies that (10) is reduced to the form

$$\frac{d}{du}\int_{x_3}^{f} u^m z^{-1}\prod_{i=1}^{4m}(z^{1/2} - x_i z^{-1/2})^{1/2}dz = 0.$$

This is the condition (8) sought.

2. The Extremal Case 2:2

Let us apply Theorem 1 to the equality case of the condition (9)/V.1.2, p. 7. Take e.g. the case $U \leqq 0$ in which the generating function $\kappa = e^{i\vartheta}$ of the equality case is defined by Theorem 1/V.1.2:

$$(11)\qquad \begin{cases} \cos\vartheta = \dfrac{\sigma}{u}, \\ \sin\vartheta = \begin{cases} \sqrt{1 - \dfrac{\sigma^2}{u^2}}, & b \leqq u \leqq c, \\ -\sqrt{1 - \dfrac{\sigma^2}{u^2}}, & c \leqq u \leqq 1; \end{cases} \\ \sigma = \left|\dfrac{U}{2\log b}\right| \in [0,b]. \end{cases}$$

In principle Löwner's equation (1) can be integrated by using the conditions (11). However, in the general unsymmetric class $S(b)$ we do not have a similar simple technique available as in the real class $S_R(b)$ (cf. I.3.9 of [1]). Therefore, the result (9) of Theorem 1 seems to be the only way of proceeding here.

Let us choose the points x_i in the case $m = 1$ so that for the continuity interval of κ

$$x_1 = x_2 = \bar{\kappa} = e^{-i\vartheta}, \quad x_3 = x_4 = \kappa = e^{i\vartheta}.$$

The conditions (2) and (4) are automatically true.

The condition (3) assumes for each x_i the form

$$u\,\frac{d\kappa}{du} = \kappa\,\frac{1 + \kappa^2}{1 - \kappa^2}$$

$\Longleftrightarrow$

$$\kappa + \kappa^{-1} + u(\kappa - \kappa^{-1})\kappa^{-1}\,\frac{d\kappa}{du} = 0$$

$\Longleftrightarrow$

$$\frac{d}{du}[u(\kappa + \kappa^{-1})] = 0.$$

This holds because (11) implies that

$$\kappa + \kappa^{-1} = 2\cos\vartheta = \frac{2\sigma}{u}.$$

Thus, the result of Theorem 1 is available. For its function F we have in the present case

$$F(z,u) = \int_{\kappa}^{f} uz^{-1}(z^{1/2} - \bar{\kappa}z^{-1/2})(z^{1/2} - \kappa z^{-1/2})dz$$

$$= \int_{\kappa}^{f} u[1 - (\kappa + \bar{\kappa})z^{-1} + z^{-2}]dz$$

$$= \Big/_{\kappa}^{f}\; u[z - z^{-1} - (\kappa + \bar{\kappa})\log z].$$

Observe, that on the lower limit

$$\frac{d}{du}[u(\kappa - \kappa^{-1}) - 2\sigma \log \kappa]$$

$$= \kappa - \kappa^{-1} + \underbrace{u(\kappa + \kappa^{-1})}_{2\sigma}\kappa^{-1}\frac{d\kappa}{du} - 2\sigma\kappa^{-1}\frac{d\kappa}{du} = \kappa - \kappa^{-1} = 2i \sin \vartheta .$$

Because

$$a_2(u) = -2\int_u^1 \cos \vartheta \, du - i\, 2\int_u^1 \sin \vartheta \, du$$

we obtain

$$\frac{d}{du} a_2(u) = 2 \cos \vartheta (u) + i\, 2 \sin \vartheta (u)$$

and hence

$$\frac{d}{du} \operatorname{Im} a_2(u) = 2 \sin \vartheta (u);$$

$$\frac{d}{du} [u(\kappa - \kappa^{-1}) - 2\sigma \log \kappa] = i \frac{d}{du} \operatorname{Im} a_2(u).$$

This shows that on an interval where κ is continuous

$$\frac{d}{du} F(z,u) = \frac{d}{du} [u(f - f^{-1}) - 2\sigma \log f - i \operatorname{Im} a_2(u)]$$

$\Rightarrow$

$$F(z,u) = u(f - f^{-1}) - 2\sigma \log f - i \operatorname{Im} a_2(u) + C.$$

Applying (9) to the intervals of (11) we deduce from this

$$b(f - f^{-1}) - 2\sigma \log f - i \operatorname{Im} a_2 = z - z^{-1} - 2\sigma \log z.$$

This is the integrated condition for the extremal function f determined by (11).

If $U > 0$, $\cos \vartheta = -\frac{\sigma}{u}$. This implies the corresponding change of sign in (12).

Theorem 2. In the equality case of the condition (9)/V.1.2., p. 7, the extremal function f is determined by

$$(12)\quad b(f - f^{-1}) \pm 2\sigma \log f - i \operatorname{Im} a_2 = z - z^{-1} \pm 2\sigma \log z.$$

$U > 0$ and the upper signs belong together as well as $U < 0$ and the lower signs.

We may apply the above result also to extremal functions 2:2 of the coefficient body (a_2, a_3) and to functions maximizing $\operatorname{Re}(a_3 + \lambda a_2)$. The examples to be considered are the following.

1°. Functions 2:2 maximizing $\operatorname{Re} \delta$ ((9)/V.1.2).

In Figure 2, p. 22, where $b = 0.1$, we choose a horizontal line segment

$$\operatorname{Im} a_2 = 1.3$$

in the second quadrant of the disc $|a_2| \leqq 2(1 - b)$. On this line segment take the points

$$\operatorname{Re} a_2 = -0.1, -0.2, -0.3, -0.4, \quad 0.2 \log 0.1 < -0.460'516.$$

In Figure 26, part 1, are presented the slits of the corresponding extremal domains determined by the connection (12) and drawn by using a computer.

2°. Boundary functions 2:2 of (a_2, a_3).

The boundary functions in question are the equality functions of (17)/V.1.2, p. 11. Consider the case $a_2 = |a_2| = U > 0$ and take $b = 0.1$, $a_2 = 1.7$ (Figure 3, the lower part, p. 25). The equation (12) applies to the rotated function

$$\begin{cases} \tilde{f}(z) = \tau^{-1} f(\tau z) = b(z + \tilde{a}_2 z^2 + \ldots), \\ \tau = e^{iv}, \\ \tilde{a}_2 = \tilde{U} + i\tilde{V}, \\ \tilde{U} = U \cos v, \quad \tilde{V} = U \sin v. \end{cases}$$

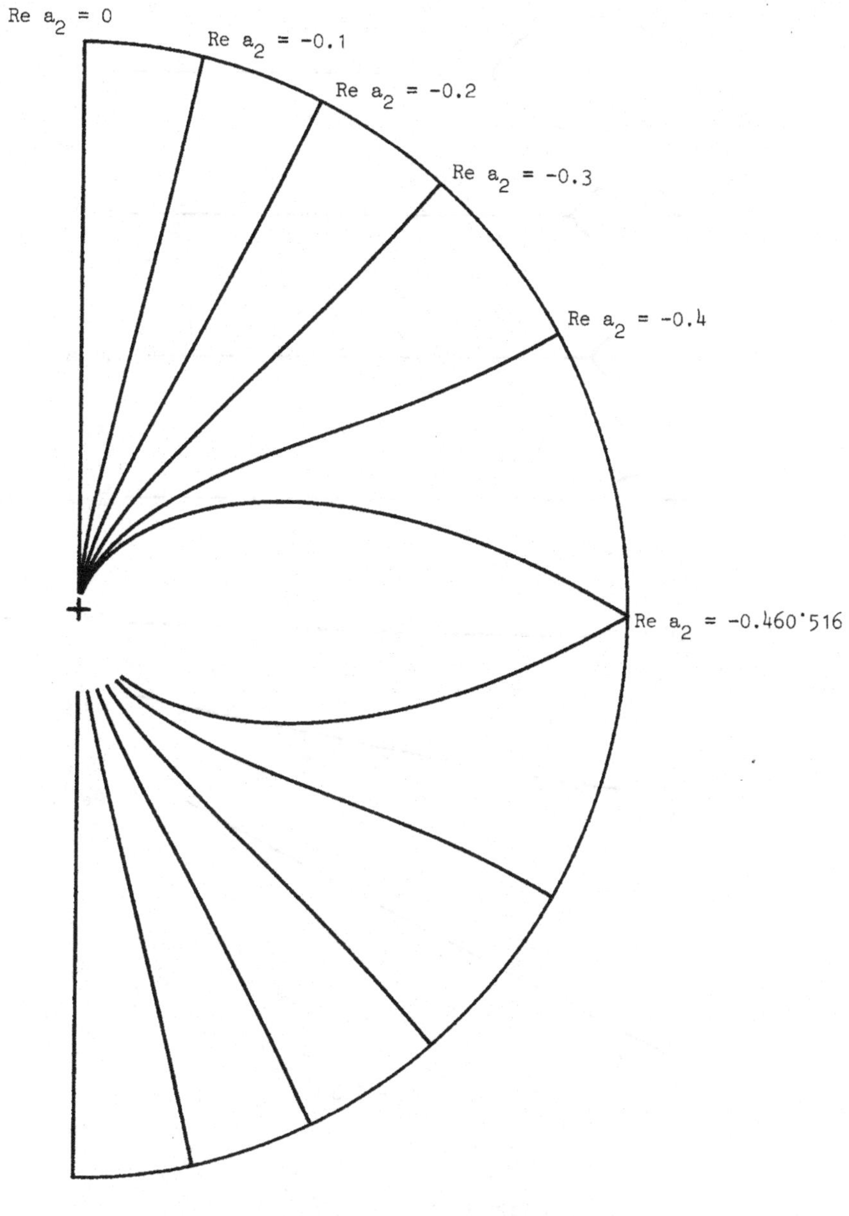

$b = 0.1$, $\mathrm{Im}\ a_2 = 1.3$

Figure 26 (part 1).

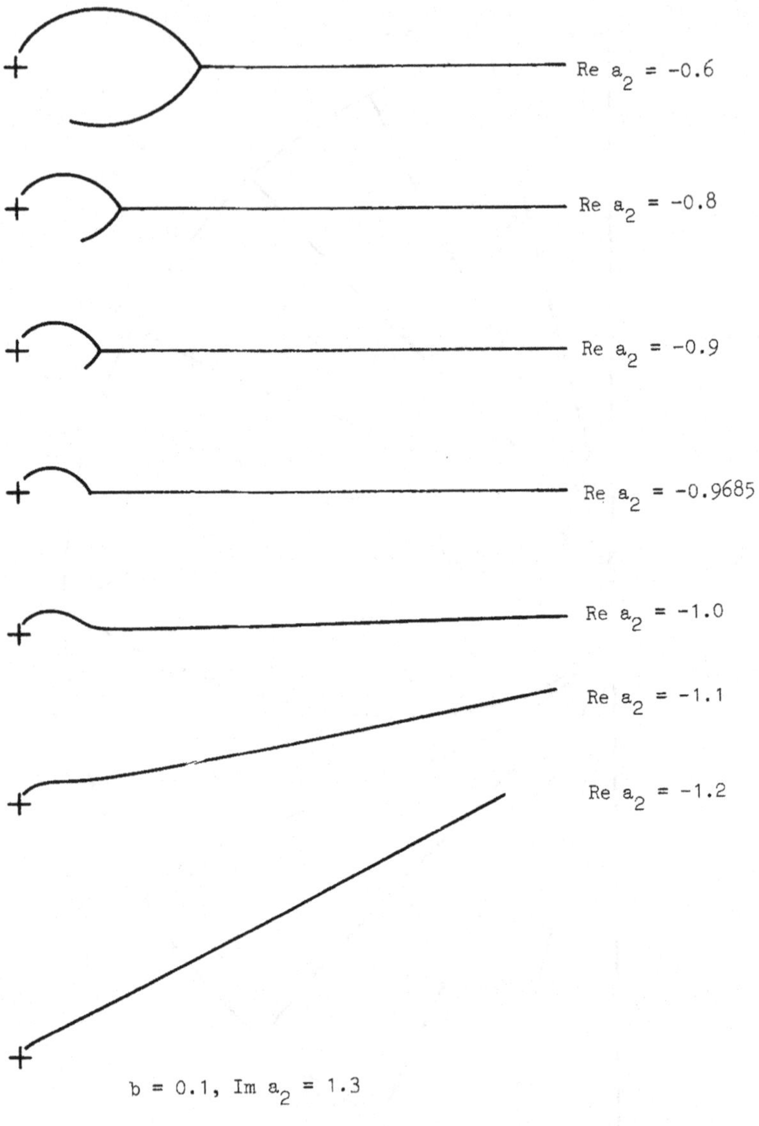

Figure 26 (part 2).

According to Table 3, p. 24, we have for v the limitation

$$0 \leqq v \leqq 1.845{\cdot}116.$$

In the present case

$$\begin{cases} \tilde{U} = 1.7 \cos v, \\ \tilde{V} = 1.7 \sin v, \\ \sigma = \left| \dfrac{\tilde{U}}{2 \log b} \right| \in [0,b], \\ b(\tilde{f} - \tilde{f}^{-1}) \pm 2\sigma \log \tilde{f} - i\tilde{V} = z - z^{-1} \pm 2\sigma \log z, \\ f(w) = \tau\tilde{f}(\tau^{-1}w) \quad (w = \tau z). \end{cases}$$

$\tilde{U} < 0$ requires the lower signs, $\tilde{U} > 0$ the upper signs.

In Figure 27, part 1, are presented the extremal domains belonging to the choices

$$v = \frac{\pi}{2}, \; 1.7, \; 1.8, \; 1.845{\cdot}116.$$

3°. Functions 2:2 maximizing Re $(a_3 + \lambda a_2)$.

Theorem 1/V.3.2, p. 66, determines the case illustrated in Figure 24, p. 92, where $b = 0.5$ and $\lambda_1 \geqq 0$, $\lambda_2 \geqq 0$. According to the results in V.3.2 we have for the extremal function f of (12):

$$\begin{cases} \sigma = \dfrac{\lambda_1}{4(1 + \log b)} \leqq b \in (e^{-1},1), \\ U = - \dfrac{\lambda_1 \log b}{2(1 + \log b)} > 0, \\ V = - \dfrac{\lambda_2}{2} < 0. \end{cases}$$

Take the vertical line segment

$$\lambda_1 = 0.5$$

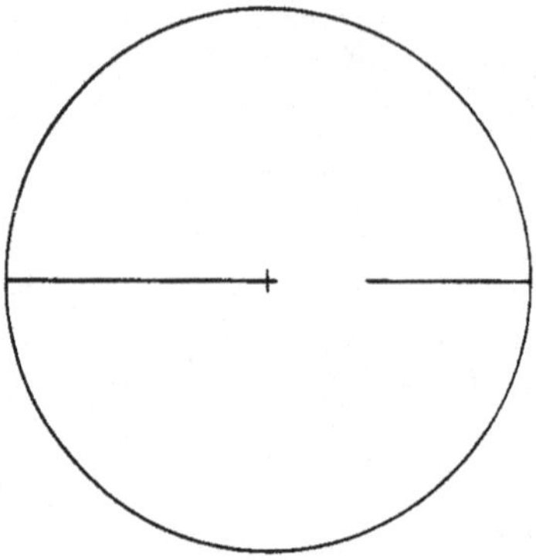

$v = \frac{\pi}{2}$

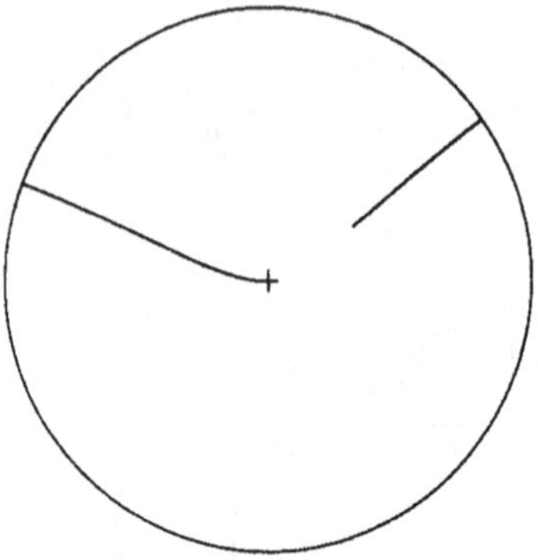

$v = 1.7$

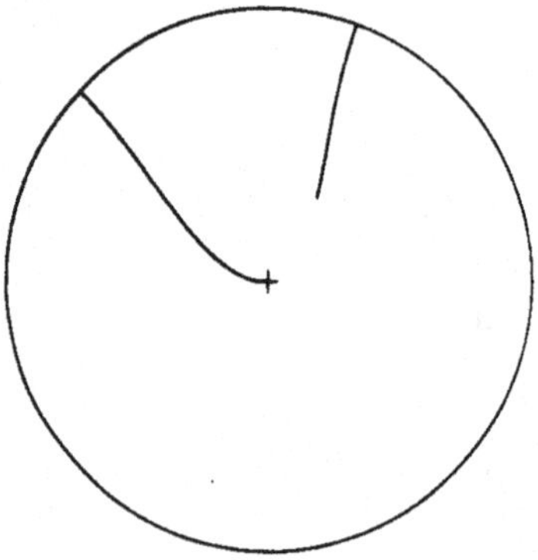

$v = 1.8$

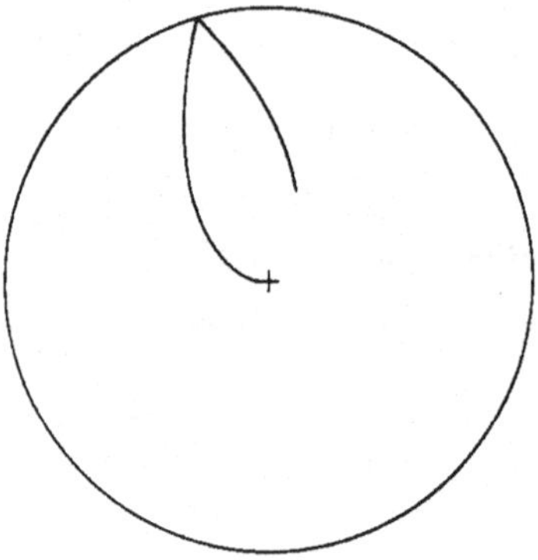

$v = 1.845{\cdot}116$

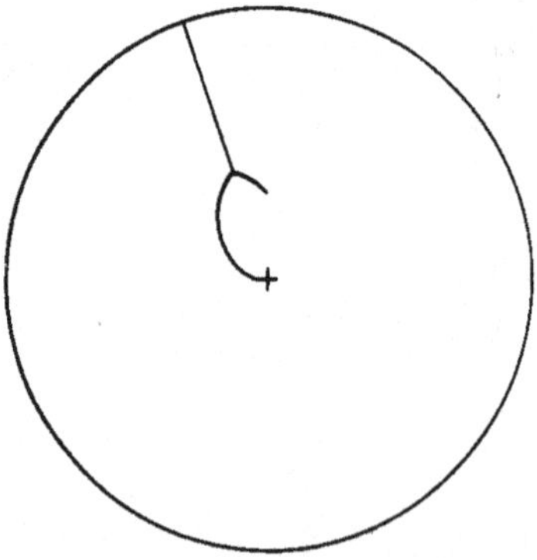

$v = 1.9$

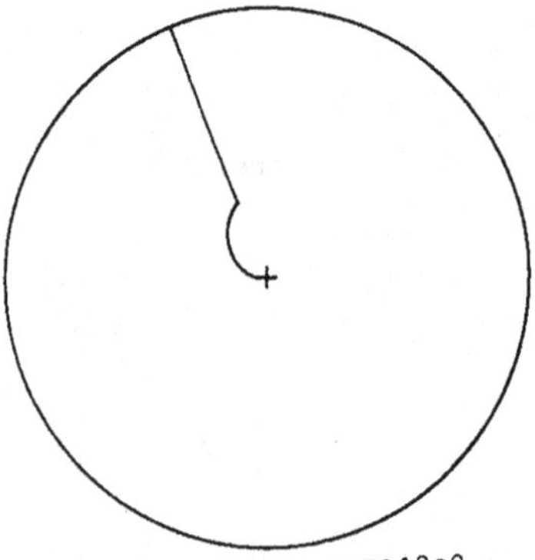

$v = 1.952{\cdot}808$

$b = 0.1,\ a_2 = 1.7$

Figure 27 (part 1).

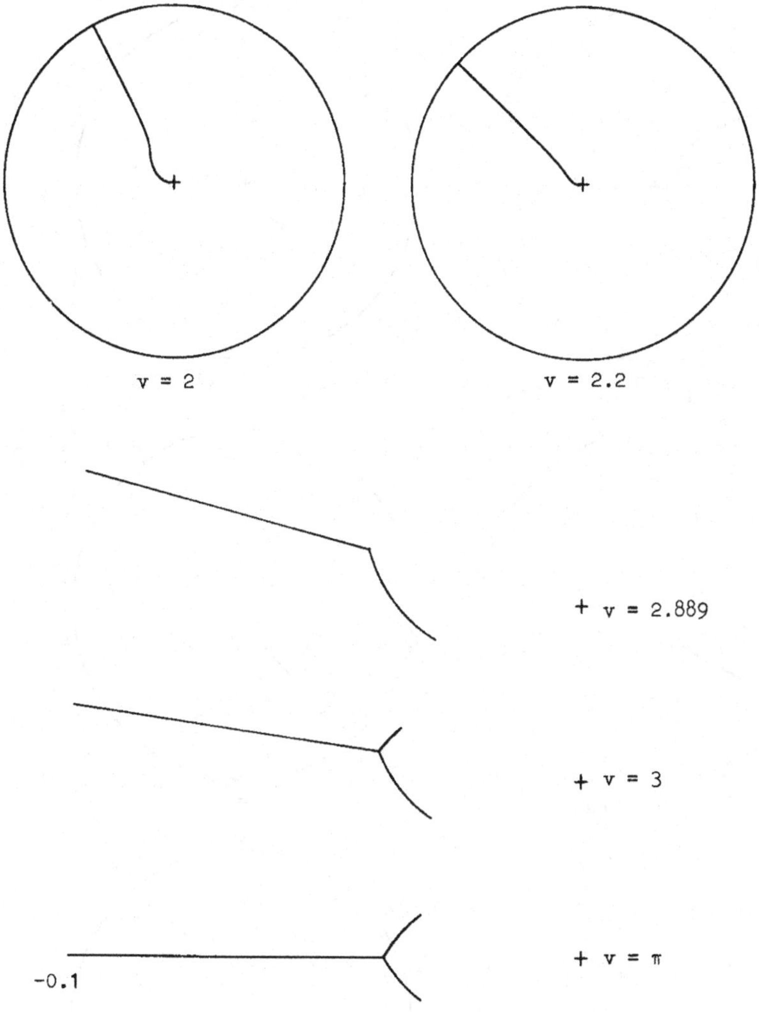

$b = 0.1$, $a_2 = 1.7$

Figure 27 (part 2).

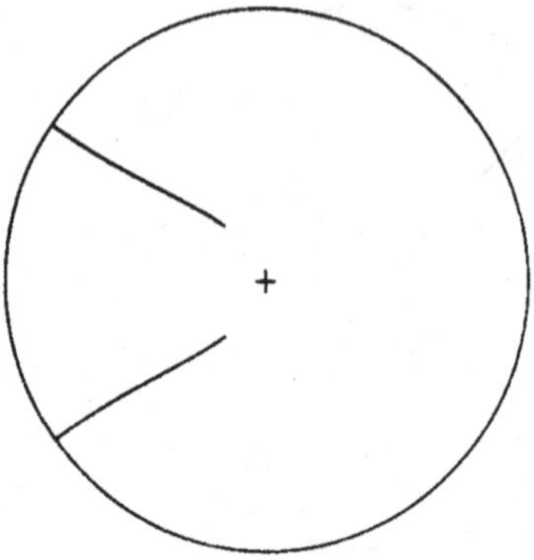

$\lambda_2 = 0$

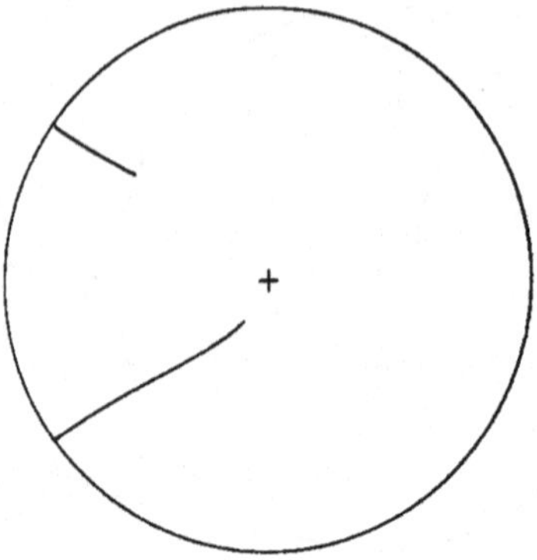

$\lambda_2 = 1.5$

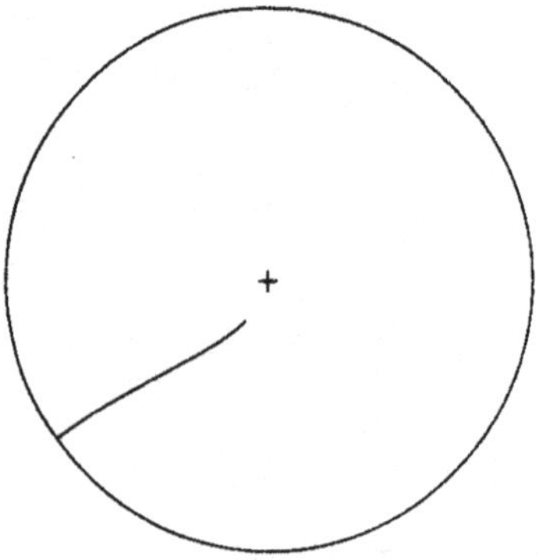

$\lambda_2 = 1.625$

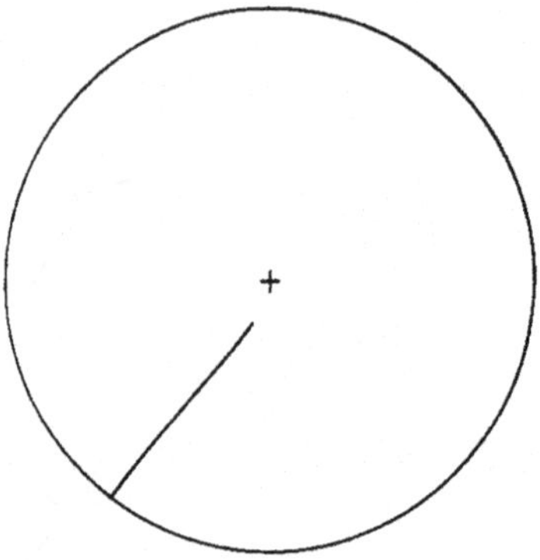

$\lambda_2 = 2$

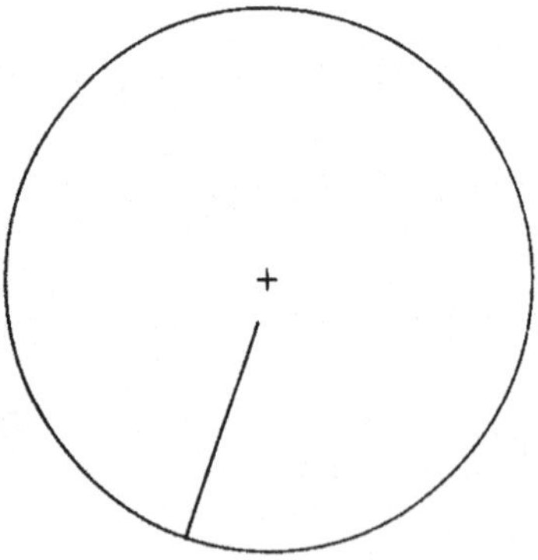

$\lambda_2 = 3$

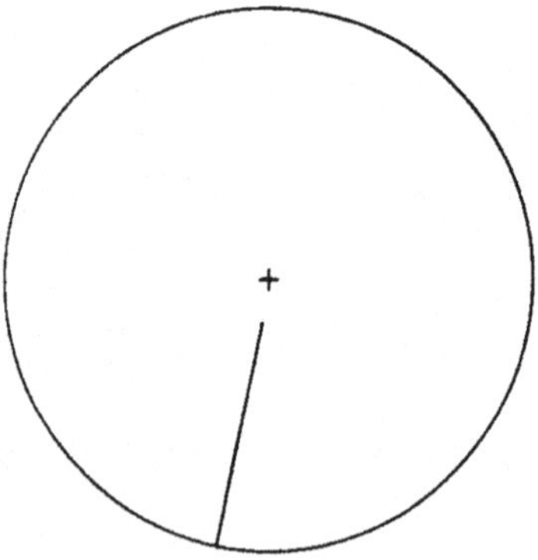

$\lambda_2 = 4$

$b = 0.5$, $\lambda_1 = 0.5$

Figure 28 (part 1).

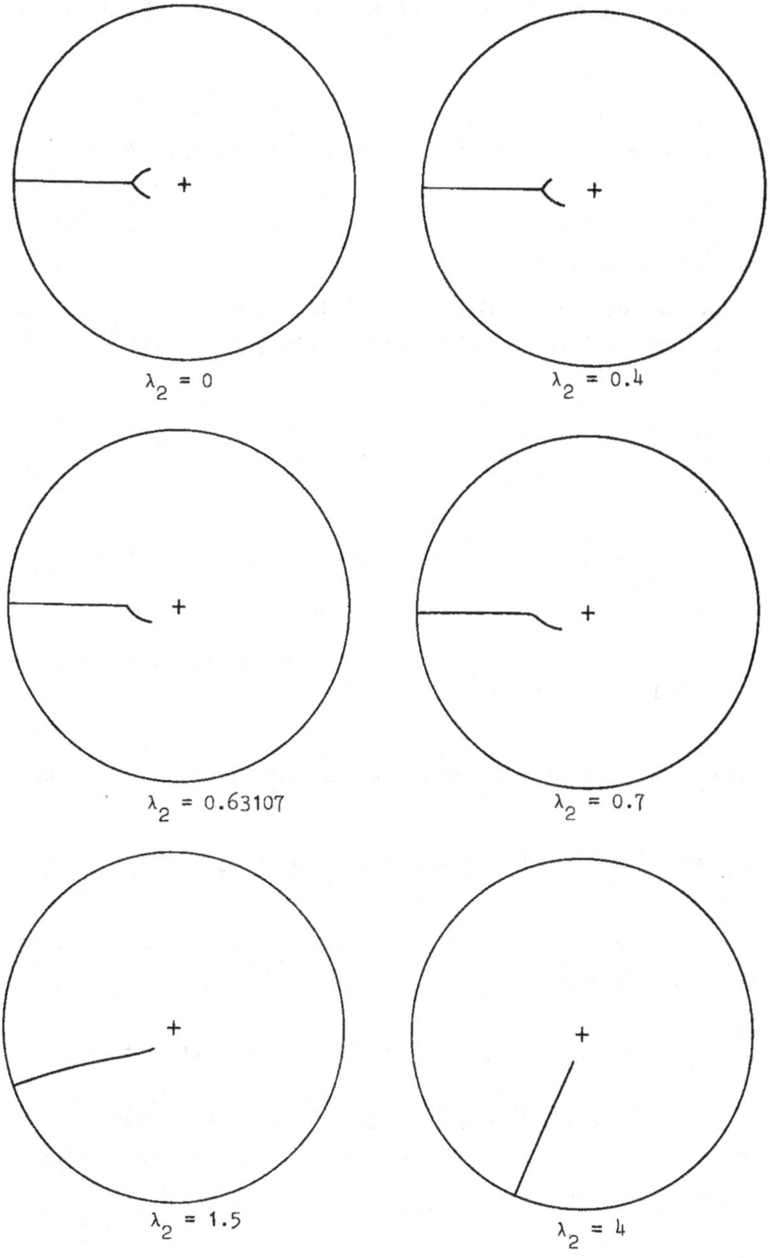

$b = 0.5,\ \lambda_1 = 1.0$

Figure 28 (part 2).

of Figure 24 and choose the following points of it (cf. Table 6, p. 79):

$$\lambda_2 = 0,\ 1.5,\ 1.625.$$

The corresponding extremal domains are presented in Figure 28, part 1.

3. The Extremal Case 1:2

Next, consider the equality case of the estimation (25)/V.1.3, p. 17. Again, let first $U \leqq 0$. In this case the equality function is generated by $\kappa = e^{i\vartheta}$ for which

$$\cos\vartheta = \begin{cases} 1, & b \leqq u \leqq \sigma, \\ \frac{\sigma}{u}, & \sigma \leqq u \leqq 1. \end{cases}$$

Here $\sigma \in [b,1]$ is the root of the equation

$$\sigma \log \sigma - \sigma + b - \frac{U}{2} = 0.$$

On the interval $[\sigma,1]$ the above result (12) is valid, giving for $f(z,\sigma)$

$$b[f(z,\sigma) - f(z,\sigma)^{-1}] - 2\sigma \log f(z,\sigma) - i \operatorname{Im} a_2(\sigma) = z - z^{-1} - 2\sigma \log z.$$

For $[b,\sigma]$ the condition $\cos\vartheta \equiv 1$ yields the right radial-slit-mapping $f(z,\sigma) \to f(z,b)$ for which

$$b[f(z,b) + f(z,b)^{-1} + 2] = \sigma[f(z,\sigma) + f(z,\sigma)^{-1} + 2].$$

These two conditions define the connection $z \to f(z,b)$ sought.

If $U > 0$ some changes of sign occur according to (27)/V.1.3.

Observe that here

$$\operatorname{Im} a_2(\sigma) = \operatorname{Im} a_2(b).$$

Theorem 3. In the equality case of the condition (25)/V.1.3, p. 17, where $\kappa = e^{i\vartheta}$ is defined by the equations (26)-(27)/V.1.3, the extremal function $f = f(z,b)$ is determined by $(a_2 = a_2(b))$

$$(13)\quad \begin{cases} b[f(z,\sigma)-f(z,\sigma)^{-1}] \pm 2\sigma \log f(z,\sigma) - i \operatorname{Im} a_2 = z - z^{-1} \pm 2\sigma \log z, \\ b(f + f^{-1} \mp 2) = \sigma[f(z,\sigma) + f(z,\sigma)^{-1} \mp 2]. \end{cases}$$

The upper signs are to be used if $U > 0$ and the lower ones for $U < 0$. Consider again examples for the above type 1°-3°.

1°. Functions 1:2 maximizing $\operatorname{Re}\delta$ ((25)/V.1.3).

Take the continuation of the previous line segment of Figure 2, p. 22:

$$\begin{cases} b = 0.1, \\ \operatorname{Im} a_2 = 1.3, \\ \operatorname{Re} a_2 = -0.6, -0.8, -0.9, -0.9685. \end{cases}$$

The corresponding extremal domains (only their slits) are computer-presented in Figure 26, part 2.

2°. Boundary functions 1:2 of (a_2,a_3).

We are again in the lower part of Figure 3, p. 25, and on the arcs determined by $v = \arg\tau$. According to Table 3, p. 24, we may choose

$$\begin{cases} v = 1.845\cdot116,\ 1.9,\ 1.952\cdot808; \\ v = 2.889,\ 3,\ \pi. \end{cases}$$

Rewrite the formulae needed for finding the extremal $f = f(z,b)$:

$$\begin{cases} \tilde U = 1.7 \cos v < 0, \\ \tilde V = 1.7 \sin v, \\ b[\tilde f(z,\sigma) - \tilde f(z,\sigma)^{-1}] - 2\sigma \log \tilde f(z,\sigma) - i\tilde V = z - z^{-1} - 2\sigma \log z, \\ b[\tilde f(z,b) - \tilde f(z,b)^{-1} + 2] = \sigma[\tilde f(z,\sigma) + \tilde f(z,\sigma)^{-1} + 2], \\ f = f(z,b) = \tau\tilde f(\tau^{-1}z,b). \end{cases}$$

The extremal domains are in Figure 27, part 1 and 2.

3°. Functions 1:2 maximizing Re $(a_3 + \lambda a_2)$.

Theorem 4/V.3.3, p. 78, gives max Re $(a_3 + \lambda a_2)$ when λ is subject to the limitation (35)/V.3.3. In Figure 24, p. 92, we take now the line segment (cf. Table 6, p. 79)

$$\begin{cases} b = 0.5, \\ \lambda_1 = 1, \\ \lambda_2 = 0,\ 0.4,\ 0.631\cdot 07. \end{cases}$$

The results of V.3.3 include the formulae needed in (13):

$$\begin{cases} \cos \vartheta = \begin{cases} -1, & b \leqq u \leqq \sigma, \\ -\frac{\sigma}{u}, & \sigma \leqq u \leqq 1; \end{cases} \\ \sigma \log \sigma + b - \frac{\lambda_1}{4} = 0, \\ U = 2\sigma - 2b - 2\sigma \log \sigma > 0, \\ V = -\frac{\lambda_2}{2} < 0. \end{cases}$$

The extremal domains are in Figure 28, part 2.

4. The Extremal Case 1:1

Finally, turn to Theorem 5/V.1.4, p. 31, where the inequality (46)/V.1.4 determines max Re δ. The extremal generating function $\kappa = e^{i\vartheta}$ can now be expressed in terms of two numbers $\alpha = -\vartheta(1)$ and $\omega = -\vartheta(b)$, the connection of which to $a_2 = U + iV$ is defined by the equations (44)-(45)/ V.1.4.

In order to find points x_i satisfying the conditions (2)-(4)/V.4.1 of Lemma 2 we introduce a function $t = t(u) \in (0,1)$ so that

$$(14) \qquad 4\sigma(u) = \frac{(t+1)^2}{t} u.$$

Here $\sigma = \sigma(u) > u$ is defined by (38)/V.1.4, p. 28. The point system x_i $(i = 1,\ldots,4)$ in question appears to be

$$(15) \qquad x_1 = x_2 = \bar{\kappa} = e^{-i\vartheta(u)}, \quad x_3 = t(u)e^{i\vartheta(u)}, \quad x_4 = \frac{1}{\bar{x}_3} = \frac{1}{t(u)} e^{i\vartheta(u)}.$$

The conditions (2) and (4)/V.4.1 are automatically true. We have to show (3)/V.4.1 for x_3. After that Lemma 1 transforms the result also for x_4.

By using the definitions (15) and (14) we determine the value of the following expression

$$u(2x_1 + x_3 + x_4) = u[2e^{-i\vartheta} + \underbrace{(t + t^{-1})}_{4\frac{\sigma}{u} - 2}e^{i\vartheta}]$$

$$= 2u(e^{-i\vartheta} + e^{i\vartheta}) + 4\sigma\, e^{i\vartheta} = 4(\sigma - u)i \sin\vartheta + 4\sigma \cos\vartheta$$

$$= 2C_1 - 2i\, C_2 = 2\bar{C}.$$

The connection to C is due to (39)/V.1.4, p. 29. Express the above $2\bar{C}$ in x_3:

$$2\bar{C} = u(2x_1 + x_3 + x_4) = u\left(2\frac{\bar{x}_3}{|x_3|} + x_3 + \frac{1}{\bar{x}_3}\right)$$

$$= u\left(2\sqrt{\frac{\bar{x}_3}{x_3}} + x_3 + \frac{1}{\bar{x}_3}\right) = u(2\bar{x}_3^{1/2}x_3^{-1/2} + x_3 + \bar{x}_3^{-1}).$$

From this a differential equation for $x_3 = x_3(u)$ follows by differentiation with respect to u:

$$0 = u(\bar{x}_3^{-1/2}x_3^{-1/2} - \bar{x}_3^{-2})\frac{d\bar{x}_3}{du} + u(-\bar{x}_3^{1/2}x_3^{-3/2} + 1)\frac{dx_3}{du}$$

$$+ 2\bar{x}_3^{1/2}x_3^{-1/2} + x_3 + \bar{x}_3^{-1}.$$

Conjugation and multiplication by $\bar{x}_3^{-1/2}x_3^{-1/2}$ yields

$$0 = \bar{x}_3^{-1/2}x_3^{-1/2}[u(x_3^{-1/2}\bar{x}_3^{-1/2} - x_3^{-2})\frac{dx_3}{du}$$

$$+ u(-x_3^{1/2}\bar{x}_3^{-3/2} + 1)\frac{d\bar{x}_3}{du} + 2x_3^{1/2}\bar{x}_3^{-1/2} + \bar{x}_3 + x_3^{-1}]$$

$$= u(-\bar{x}_3^{-2} + \bar{x}_3^{-1/2}x_3^{-1/2})\frac{d\bar{x}_3}{du}$$

$$+ u(x_3^{-1}\bar{x}_3^{-1} - x_3^{-5/2}\bar{x}_3^{-1/2})\frac{dx_3}{du} + 2\bar{x}_3^{-1} + x_3^{-1/2}\bar{x}_3^{1/2} + x_3^{-3/2}\bar{x}_3^{-1/2}.$$

Subtract the two equations:

$$u(1 + x_3^{-5/2}\bar{x}_3^{-1/2} - x_3^{-3/2}\bar{x}_3^{1/2} - x_3^{-1}\bar{x}_3^{-1})\frac{dx_3}{du}$$

$$+ x_3^{-1/2}\bar{x}_3^{-1/2} - \bar{x}_3^{-1} + x_3 - x_3^{-3/2}\bar{x}_3^{-1/2} = 0$$

$\Rightarrow$

$$u(1 - \bar{x}_3^{-1}x_3^{-1})(1 - \bar{x}_3^{1/2}x_3^{-3/2})\frac{dx_3}{du} + (x_3 - \bar{x}_3^{-1})(1 + \bar{x}_3^{1/2}x_3^{-3/2}) = 0.$$

Because $|x_3| = t < 1$, we obtain from this

$$u\,\frac{dx_3}{du} = -x_3\,\frac{1 + \bar{x}_3^{1/2}x_3^{-3/2}}{1 - \bar{x}_3^{1/2}x_3^{-3/2}} = x_3\,\frac{1 + \bar{x}_3^{-1/2}x_3^{1/2}\cdot x_3}{1 - \bar{x}_3^{-1/2}x_3^{1/2}\cdot x_3} = x_3\,\frac{1 + \kappa x_3}{1 - \kappa x_3}.$$

Here $\bar{x}_3^{-1/2}x_3^{1/2} = e^{i\vartheta} = \kappa$. Thus it is proved that (3)/V.4.1 holds for x_3 and hence for x_4 too.

The function $F(z,u)$ of Theorem 1/V.4.1 is in the present case

$$F(z,u) = \int_{x_3}^{f(z,u)} uz^{-1}(z^{1/2}-x_1z^{-1/2})^{1/2}(z^{1/2}-x_2z^{-1/2})^{1/2}(z^{1/2}-x_3z^{-1/2})^{1/2}(z^{1/2}-x_4z^{-1/2})^{1/2}$$

$$= u\int_{x_3}^{f(z,u)} \frac{(z - x_1)\sqrt{(z - x_3)(z - \bar{x}_3^{-1})}}{z^2}\,dz$$

and thus, according to (9)/V.4.1 of this Theorem

$$F(f,b) = F(z,1),$$

where $f = f(z,b)$, $f(z,1) = z$.

The parameters left in this equation are listed in (16)-(17) and illustrated in Figure 29.

$$(16)\qquad \begin{cases} x_1(b) = \bar{\kappa}(b) = e^{-i\vartheta(b)} = e^{i\omega}, \\ x_3(b) = t(b)e^{i\vartheta(b)} = \tau e^{-i\omega}, \\ \tau = t(b); \end{cases}$$

$$(17)\qquad \begin{cases} x_1(1) = \bar{\kappa}(1) = e^{-i\vartheta(1)} = e^{i\alpha}, \\ x_3(1) = t(1)e^{i\vartheta(1)} = \rho e^{-i\alpha}, \\ \rho = t(1). \end{cases}$$

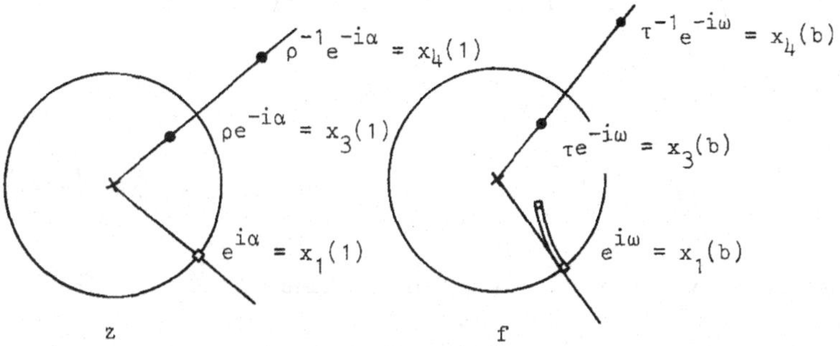

Figure 29.

<u>Theorem 4.</u> In the equality case of the condition (46)/V.1.4, p. 32, $\kappa = e^{i\vartheta}$ is determined by the parameters $\alpha = -\vartheta(1)$ and $\omega = -\vartheta(b)$, defined by (44)-(45)/V.1.4 and by aid of (39)/V.1.4. For the extremal function $f = f(z,b)$ there holds Schiffer's equation

$$(18)\quad b\int_{x_3(b)}^{f}\frac{(f-x_1(b))\sqrt{(f-x_3(b))(f-\bar{x}_3(b)^{-1})}}{f^2}\,df$$

$$=\int_{x_3(1)}^{z}\frac{(z-x_1(1)\sqrt{(z-x_3(1))(z-\bar{x}_3(1)^{-1})}}{z^2}\,dz.$$

The parameters $x_1(b)$, $x_3(b)$, $x_1(1)$, $x_3(1)$ are expressed in α and ω according to (16)-(17). Expecially the numbers $\tau = t(b)$ and $\rho = t(1)$ are obtained in α and ω by using the connections (14) and (39)/V.1.4, p. 29.

For numerical purposes the equation (18) can be integrated in the following algebraic form

$$(19)\qquad b\Big/_{x=x_3(b)}^{x=f}\ [K(x,t(b),\kappa(b)) - \overline{K((\bar{x})^{-1},t(b),x(b))}]$$

$$=\Big/_{x=x_3(1)}^{x=z}\ [K(x,t(1),\kappa(1)) - \overline{K((\bar{x})^{-1},t(1),\kappa(1))}],$$

where

$$(20)\quad K(z,t,\kappa) = \sqrt{x-t\kappa}\sqrt{x-\tfrac{1}{t}\kappa} - (t\kappa + \tfrac{1}{t}\kappa + 2\bar{\kappa})\log(\sqrt{x-tx}+\sqrt{x-\tfrac{1}{t}\kappa}).$$

As before, we apply the results in examples of the type 1°-3°.

1°. Functions 1:1 maximizing Re (δ) ((46)/V.1.4).

Consider the final continuation of the line segment of Figure 2, p. 22:

$$\begin{cases} b = 0.1, \\ \operatorname{Im} a_2 = 1.3, \\ \operatorname{Re} a_2 = -1.0,\ -1.1,\ -1.2. \end{cases}$$

The slits of the corresponding extremal domains are in Figure 26, part 2.

2°. Boundary functions 1:1 of (a_2,a_3).

We consider the lower part of Figure 3, p. 25, and there the remaining arcs determined by Table 4, p. 47. Let us choose

$$v = 2,\ 2.2,\ 2.889.$$

We take

$$\tilde{U} = 1.7 \cos v < 0, \quad \tilde{V} = 1.7 \sin v > 0$$

and apply these numbers in determining α and ω (Theorem 9/V.1.4, p. 45) needed for finding $\tilde{f}$ from Theorem 4. The extremal function sought is finally

$$f(z,b) = \tau\tilde{f}(\tau^{-1}z,b).$$

The corresponding extremal domains are in Figure 27, part 2.

3°. Functions 1:1 maximizing Re $(a_3 + \lambda a_2)$.

Theorem 5/V.3.4, p. 68, concerns max Re $(a_3 + \lambda a_2)$ when λ lies in the complement of the previous cases. In Figure 24, p. 92, we may proceed along the line segments studied before, i.e. we choose

$$\begin{cases} b = 0.5, \\ \lambda_1 = 0.5, \\ \lambda_2 = 2,\ 3,\ 4; \end{cases} \qquad \begin{cases} b = 0.5, \\ \lambda_1 = 1, \\ \lambda = 0.7,\ 1.5,\ 4. \end{cases}$$

The corresponding initial values α and ω needed are, according to Theorem 5/V.3.4, p. 86, determined by (54)/V.3.4. When applying these numbers in Theorem 4 we obtain the extremal function f sought. The extremal domains are in Figure 28, part 1 and 2.

VI ON THE SECOND COEFFICIENT BODY IN $S_R(b)$

1 § A Löwner-identity for Finding the Power-inequality for a Second Coefficient Body in $S_R(b)$

1. Löwner-identities

In [1] the Power-inequality was used in acquiring information about the second coefficient body, to be denoted here by (a_2,a_3,a_4) (III.4, pp. 266-286). In what follows we are going to rederive the results for the $S_R(b)$-functions by using the Power-inequality which will be rederived from a proper Löwner-identity. This method allows parametrizing the equality functions by aid of parameters connected with the slit-character of the extremal domains. Further, the previous results will be extended by generalizing first the Power-inequality properly. The method, based on another Löwner-identity, is the one suggested in Chapter IV. It was used above in Chapter V.

In $S_R(b)$ the following Löwner-formulae ([1], (15)/I.3.8, p. 70) hold:

$$(1)\qquad \begin{cases} a_2 = -2\displaystyle\int_b^1 \cos\vartheta\, du, \\ a_3 = a_2^2 - 2\displaystyle\int_b^1 u\cos 2\vartheta\, du, \\ a_4 = 3a_2a_3 - 2a_2^3 - 4\displaystyle\int_b^1 u\cos 2\vartheta\left(\int_b^u \cos\vartheta\, du_1\right)du - 2\displaystyle\int_b^1 u^2\cos 3\vartheta\, du. \end{cases}$$

The generating function ϑ is piecewise continuous on $[b,1]$. Rewrite the double integral:

$$\int_b^1 u\cos 2\vartheta\left(\int_b^u \cos\vartheta\, du_1\right)du = \int_b^1\int_b^u \cos\vartheta\, du_1\, d\int_1^u u_1\cos 2\vartheta\, du_1$$

$$= \Bigg/_{\!\!b}^{1}\int_1^u u_1\cos 2\vartheta\, du_1 \cdot \int_b^u \cos\vartheta\, du_1 - \int_b^1 \cos\vartheta\left(\int_1^u u_1\cos 2\vartheta\, du_1\right)du$$

$$= -\int_b^1 \int_1^u u \cos 2\vartheta \, du_1 \, d\int_1^u \cos\vartheta \, du_1$$

$$= -\int_b^1 \int_1^u \cos\vartheta \, du_1 \cdot \int_1^u u_1 \cos 2\vartheta \, du_1 + \int_b^1 u \cos 2\vartheta \left(\int_1^u \cos\vartheta \, du_1\right) du$$

$$= \underbrace{\int_b^1 \cos\vartheta \, du}_{-\frac{1}{2}a_2} \cdot \underbrace{\int_b^1 u \cos 2\vartheta \, du}_{-\frac{1}{2}(a_3 - a^2)} - \int_b^1 u \cos 2\vartheta \left(\int_u^1 \cos\vartheta \, du_1\right) du$$

$$= \frac{1}{4} a_2 a_3 - \frac{1}{4} a_2^3 - \int_b^1 u \cos 2\vartheta \left(\int_u^1 \cos\vartheta \, du_1\right) du$$

⇒

(2) $$a_4 = 2a_2a_3 - a_2^3 - 4\int_b^1 u \cos 2\vartheta \left(\int_1^u \cos\vartheta \, du_1\right) du - 2\int_b^1 u^2 \cos 3\vartheta \, du.$$

Introduce the combinations

(3) $$\begin{cases} \delta_1 = a_3 - \frac{3}{4} a_2^2, \\ \delta_2 = a_4 - 2a_2a_3 + \frac{13}{12} a_2^3. \end{cases}$$

In order to rewrite δ_2 properly we observe that

$$\int_b^1 \cos\vartheta \left(\int_1^u \cos\vartheta \, du_1\right)^2 du = \int_b^1 \left(\int_1^u \cos\vartheta \, du_1\right)^2 d\int_1^u \cos\vartheta \, du_1$$

$$= \int_1^b \frac{1}{3}\left(\int_1^u \cos\vartheta \, du_1\right)^3 = \frac{1}{3}\left(\int_b^1 \cos\vartheta \, du\right)^3$$

⇒

$$(4)\qquad a_2^3 = -8\left(\int_b^1 \cos\vartheta\, du_1\right)^3 = -24\int_b^1 \cos\vartheta\left(\int_u^1 \cos\vartheta\, du_1\right)^2 du.$$

Thus

$$(5)\qquad \delta_2 = -2\int_b^1 \cos\vartheta\left(\int_u^1 \cos\vartheta\, du_1\right)^2 du + 4\int_b^1 u\cos 2\vartheta\left(\int_u^1 \cos\vartheta\, du_1\right) - 2\int_b^1 u^2\cos 3\vartheta.$$

It appears, that the following perfect square can be combined with δ_2:

$$(6)\qquad \int_b^1\left(\cos\frac{\vartheta}{2}\int_u^1 \cos\vartheta\, du_1 - u\cos\frac{3\vartheta}{2}\right)^2 du = \int_b^1 A^2 du;$$

$$A = \cos\frac{\vartheta}{2}\int_u^1 \cos\vartheta\, du_1 - u\cos\frac{3\vartheta}{2}.$$

For A^2 we have

$$A^2 = \underbrace{\cos^2\frac{\vartheta}{2}}_{\frac{1+\cos\vartheta}{2}}\left(\int_u^1 \cos\vartheta\, du_1\right)^2 - 2u\underbrace{\cos\frac{\vartheta}{2}\cos\frac{3\vartheta}{2}}_{\cos\vartheta + \cos 2\vartheta}\int_u^1 \cos\vartheta\, du_1 + u^2\underbrace{\cos^2\frac{3\vartheta}{2}}_{\frac{1+\cos 3\vartheta}{2}}$$

$$= \frac{1}{2}\left(\int_u^1 \cos\vartheta\, du_1\right)^2\frac{du}{du} + \frac{d}{du}\left[\frac{1}{2}\left(\int_u^1 \cos\vartheta\, du_1\right)^2\right]u - \frac{d}{du}\frac{1}{6}\left(\int_u^1 \cos\vartheta\, du_1\right)^3$$

$$- u\cos 2\vartheta\int_u^1 \cos\vartheta\, du_1 + \frac{u^2}{2} + \frac{u^2}{2}\cos 3\vartheta$$

⇒

$$\int_b^1 A^2 du = \int_b^1 \left[\frac{u}{2}\left(\int_u^1 \cos\vartheta\, du_1\right)^2 - \frac{1}{6}\left(\int_u^1 \cos\vartheta\, du_1\right)^3 + \frac{u^3}{6}\right]$$

$$+ \underbrace{\int_b^1 u \cos 2\vartheta \left(\int_1^u \cos\vartheta\, du_1\right) du + \frac{1}{2}\int_b^1 u^2 \cos 3\vartheta\, du}_{-\frac{a_4 - 2a_2a_3 + a_2^3}{4}} \quad \text{(from (2))}$$

$$= -\frac{b}{2}\underbrace{\left(\int_b^1 \cos\vartheta\, du\right)^2}_{\frac{a_2^2}{4}} + \frac{1}{6}\underbrace{\left(\int_b^1 \cos\vartheta\, du\right)^3}_{-\frac{a_2^3}{8}} + \frac{1}{6}(1 - b^3) - \frac{a_4 - 2a_2a_3 + a_2^3}{4}$$

⇒

(7) $$\delta_2 + \frac{b}{2} a_2^2 - \frac{2}{3}(1 - b^3) = -4 \int_b^1 A^2 du.$$

By aid of this identity one can finally prove that

(8) $$\delta_2 + \frac{b}{2} a_2^2 - \frac{2}{3}(1 - b^3) + 2\lambda(\delta_1 + ba_2) + \lambda^2(a_2 - 2(1-b)) = -4 \int_b^1 \cos^2 \frac{\vartheta}{2} K(u)^2 du,$$

(9) $$K(u) = \int_u^1 \cos\vartheta\, du_1 - u(2\cos\vartheta - 1) - \lambda, \quad \lambda \in R.$$

In order to check (8) observe that

$$\cos\frac{\vartheta}{2} K(u) = A - \lambda \cos\frac{\vartheta}{2}$$

and hence the right side of (8) assumes the form

$$(10)\quad -4\int_b^1 A^2 du + 8\lambda \int_b^1 \left(\cos^2\frac{\vartheta}{2}\int_u^1 \cos\vartheta\, du_1 - u\cos\frac{\vartheta}{2}\cos\frac{3\vartheta}{2}\right)du - 4\lambda^2 \int_b^1 \cos^2\frac{\vartheta}{2}$$

Consider the two last integrals of this expression.

$$\int_b^1 \left(\underbrace{\cos^2\frac{\vartheta}{2}}_{\frac{1+\cos\vartheta}{2}}\int_u^1 \cos\vartheta\, du_1 - u\underbrace{\cos\frac{\vartheta}{2}\cos\frac{3\vartheta}{2}}_{\frac{\cos\vartheta+\cos 2\vartheta}{2}}\right)du$$

$$= \frac{1}{2}\int_b^1 \left(\underbrace{\left(\int_u^1 \cos\vartheta\, du_1 - u\cos\vartheta\right.}_{\frac{d}{du}\left(u\int_u^1 \cos\vartheta\, du_1\right)} + \cos\vartheta\int_u^1 \cos\vartheta\, du_1 - u\cos 2\vartheta\right)du$$

$$= \frac{1}{2}\Big/_b^1\, u\int_u^1 \cos\vartheta\, du_1 - \frac{1}{4}\Big/_b^1\left(\int_u^1 \cos\vartheta\, du_1\right)^2 - \frac{1}{2}\int_b^1 u\cos 2\vartheta\, du$$

$$= -\frac{b}{2}\underbrace{\int_b^1 \cos\vartheta\, du}_{-\frac{a_2}{2}} + \frac{1}{4}\underbrace{\left(\int_b^1 \cos\vartheta\, du\right)^2}_{\frac{a_2^2}{4}} - \frac{1}{2}\underbrace{\int_b^1 u\cos 2\vartheta\, du}_{-\frac{a_3-a_2^2}{2}}$$

$\Rightarrow$

$$(11)\quad \int_b^1 \left(\cos^2\frac{\vartheta}{2}\int_u^1 \cos\vartheta\, du_1 - u\cos\frac{\vartheta}{2}\cos\frac{3\vartheta}{2}\right)du = \frac{b}{4}a_2 + \frac{a_2^2}{16} + \frac{a_3 - a_2^2}{4}.$$

Further,

$$(12)\quad \int_b^1 \cos^2\frac{\vartheta}{2}\, du = \int_b^1 \frac{1+\cos\vartheta}{2}\, du = \frac{1}{2}(1-b) + \frac{1}{2}\int_b^1 \cos\vartheta\, du = \frac{1}{2}(1-b) - \frac{a_2}{4}.$$

The identity (8) can now be verified by determining the value of its right side from (10) by aid of (12), (11) and (7).

The identity (8) implies the Power-inequality in $S_R(b)$ for a_4. It is especially useful for parametrizing the equality cases. It was found by O. Jokinen who also discovered a more general identity good for the S(b)-functions [6].

2. The Corresponding Power-inequality and the Equality Functions.

The condition (8) holds identically for all piecewise continuous functions ϑ. Thus, because the right side vanishes for

$$\cos \frac{\vartheta}{2} \equiv 0$$

the same holds for the left side. This can be checked directly for the left radial-slit-mapping for which $\vartheta \equiv \pi$ and

$$\begin{cases} a_2 = 2(1 - b) \\ a_3 = 3 - 8b + 5b^2, \\ a_4 = 4 - 20b + 30b^2 - 14b^3. \end{cases}$$

In what follows, the left radial-slit-mapping will be excluded. The coefficient of λ^2 in (8) is thus < 0.

Consider those general equality cases where

$$(13) \qquad \begin{cases} \cos \dfrac{\vartheta}{2} \equiv 0, & b \leqq u \leqq \sigma \\ K(u) \equiv 0, & \sigma \leqq u \leqq 1. \end{cases}$$

From the latter condition it follows that for a differentiable ϑ necessarily

$$\frac{d}{du} K(u) = -3 \cos \vartheta + 1 + 2u \sin \vartheta \, \frac{d\vartheta}{du} = 0$$

$\Rightarrow$

$$u \frac{d\vartheta}{du} = \frac{3 \cos \vartheta - 1}{2 \sin \vartheta}$$

$\Rightarrow$

$$\cos \vartheta = \frac{1}{3} + Cu^{-3/2} .$$

The constant C is determined by substituting this family of candidates in the original condition $K(u) = 0$:

$$K(u) = \int_u^1 \left(\frac{1}{3} + Cu^{-3/2}\right)du_1 - u\left(\frac{2}{3} + 2Cu^{-3/2} - 1\right) - \lambda$$

$$= \frac{1}{3} - 2C - \lambda = 0$$

$\Rightarrow$

$$C = \frac{1 - 3\lambda}{6};$$

(14) $$\cos\vartheta = \frac{1}{3} + \frac{1 - 3\lambda}{6} u^{-3/2}, \quad b \leqq \sigma \leqq u \leqq 1.$$

This result leads us to two main equality cases. Observe, that in a way typical to Löwner-presentation, there exists an infinite number of ways for combining the two conditions $\cos\frac{\vartheta}{2} \equiv 0$ and $K(u) \equiv 0$. However, there remain only two essentially different cases, to be studied next.

1°. $1 - 3\lambda \geqq 0 \Leftrightarrow \lambda \leqq 1/3$.

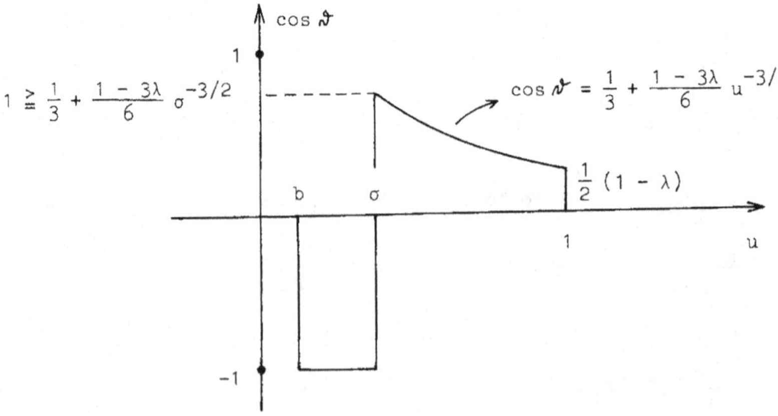

Figure 30.

The condition $\cos\vartheta \in [-1,1]$ requires that

$$-1 \leqq \frac{1}{3} + \frac{1-3\lambda}{6} = \frac{1}{2}(1-\lambda) \leqq \frac{1}{3} + \frac{1-3\lambda}{6}\sigma^{-3/2} \leqq 1.$$

The left inequality is equivalent to $\lambda \leqq 3$ and thus holds automatically. The right inequality implies that

$$0 \leqq 1 - 3\lambda \leqq 4\sigma^{3/2}$$

$\Longleftrightarrow$

(15) $$\frac{1}{3} - \frac{4}{3}\sigma^{3/2} \leqq \lambda \leqq \frac{1}{3}.$$

2°. $1 - 3\lambda \leqq 0 \Longleftrightarrow \lambda \geqq \frac{1}{3}$.

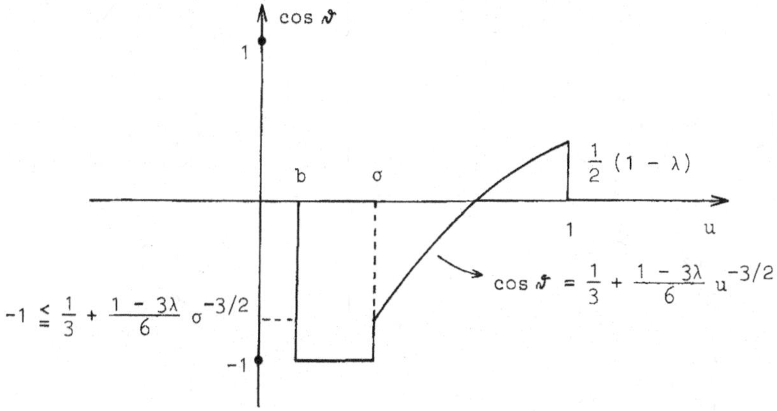

Figure 31.

As before we have

$$-1 \leqq \frac{1}{3} + \frac{1-3\lambda}{6}\sigma^{-3/2} \leqq \frac{1}{3} + \frac{1-3\lambda}{6} = \frac{1}{2}(1-\lambda) \leqq 1.$$

The right inequality is equivalent to $\lambda \geqq -1$, which condition holds automatically, and the left inequality implies that

(16) $$\frac{1}{3} \leqq \lambda \leqq \frac{1}{3} + \frac{8}{3}\sigma^{3/2}.$$

In Figure 32 there is a schematic presentation of the types of image domains connected with the equality functions when σ and λ are used as parameters.

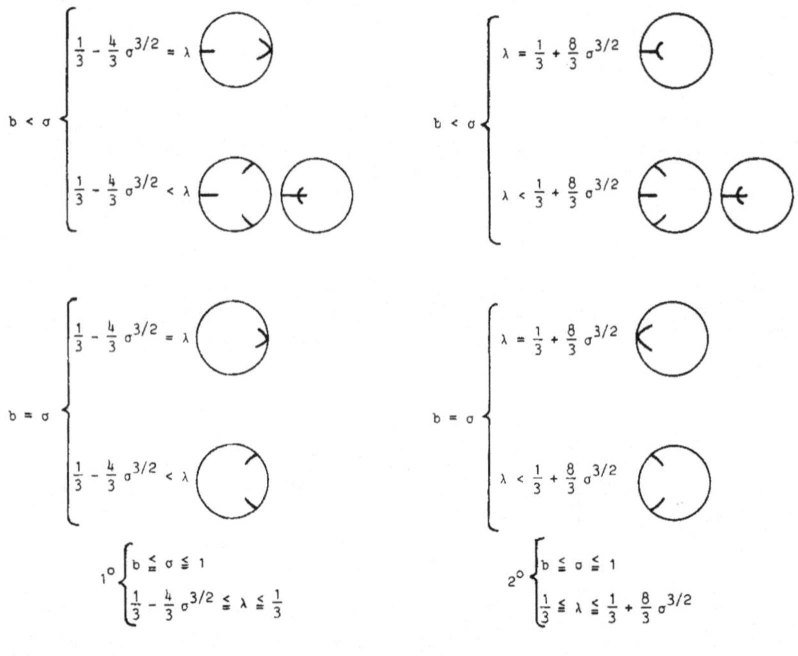

Figure 32.

Rewrite the identity (8) by using abbreviations:

(17) $$\delta_2 = \underbrace{\overbrace{\frac{2}{3}(1-b^3) - \frac{b}{2}a_2^2}^{A} - \overbrace{2\lambda(\delta_1+ba_2)}^{B} + \overbrace{(2(1-b)-a_2)}^{C>0}\lambda^2}_{I} - \underbrace{4\int_b^1 \cos^2\frac{\vartheta}{2}K(u)^2du}_{II};$$

$$\begin{cases} A = \frac{2}{3}(1 - b^3) - \frac{b}{2} a_2^2, \\ B = \delta_1 + b a_2, \\ C = 2(1 - b) - a_2 > 0. \end{cases}$$

We can estimate II upwards:

$$\text{II} \leqq 0$$

and we know the equality function ϑ. In either case 1° and 2° there holds:

$$(18) \qquad \begin{cases} \cos\vartheta = \begin{cases} -1 & , \quad b \leqq u < \sigma, \\ \frac{1}{3} + ku^{-3/2}, & \sigma \leqq u \leqq 1; \end{cases} \\ k = \frac{1 - 3\lambda}{6}, \\ \frac{1}{3} - \frac{4}{3}\sigma^{3/2} \leqq \lambda \leqq \frac{1}{3} + \frac{8}{3}\sigma^{3/2}. \end{cases}$$

This estimation yields the Power-inequality for the $S_R(b)$-functions:

$$(19) \qquad \delta_2 \leqq A - 2B\lambda + C\lambda^2 = C(\lambda - \frac{B}{C})^2 + \frac{AC - B^2}{C} = F(\lambda),$$

which thus can be optimized in the form

$$(20) \qquad \delta_2 \leqq \frac{AC - B^2}{C} = F(\frac{B}{C})$$

by choosing

$$(21) \qquad \lambda = \frac{B}{C}.$$

Now the following question arises: Is the choice (21) of λ possible for the equality function (18)? If so, then the optimized inequality (20) is sharp and has also the equality function (18). In order to check this we have to determine the coefficients a_2 and a_3, by aid of (18), from (1).

$$-\frac{1}{2}a_2 = \int_b^1 \cos\vartheta\,du = \int_b^\sigma (-1)du + \int_\sigma^1 (\tfrac{1}{3} + ku^{-3/2})du$$

$$= b - \sigma + \frac{1}{3}(1 - \sigma) - 2k\left(1 - \frac{1}{\sqrt{\sigma}}\right);$$

$$a_2 = 2(\sigma - b) - \frac{2}{3}(1 - \sigma) + 4k\left(1 - \frac{1}{\sqrt{\sigma}}\right).$$

$$\delta_2 = a_3 - a_2^2 = -2\int_b^1 u(2\cos^2\vartheta - 1)du = 1 - b^2 - 4\int_b^1 u\cos^2\vartheta\,du$$

$\Rightarrow$

$$-\frac{1}{4}[a_3 - a_2^2 - (1-b^2)] = \int_b^1 u\cos^2\vartheta\,du = \int_b^\sigma udu + \int_\sigma^1 u(\tfrac{1}{9} + \tfrac{2}{3}ku^{-3/2} + k^2u^{-3})du$$

$$= \frac{1}{2}(\sigma^2-b^2) + \frac{1}{18}(1-\sigma^2) + \frac{4}{3}k(1-\sqrt{\sigma}) + k^2\,\frac{1-\sigma}{\sigma};$$

$$a_3 = a_2^2 + \underbrace{1 - b^2 - 2(\sigma^2-b^2) - \frac{2}{9}(1-\sigma^2) - \frac{16}{3}k(1-\sqrt{\sigma}) - 4k^2\frac{1-\sigma}{\sigma}}_{h}.$$

$$\frac{B}{C} = \frac{a_3 - \frac{3}{4}a_2^2 + ba_2}{2(1-b) - a_2} = \frac{\frac{a_2^2}{4} + ba_2 + h}{2(1-b) - a_2}.$$

$$\frac{a_2^2}{4} + ba_2 = \frac{a_2^2}{4} + ba_2 + (-a_2^2+b^2-2b+1) + (a_2^2-b^2+2b-1) + 2 - 2b - 2 + 2b$$

$$= (\frac{a_2}{2} + b - 1)^2 + a_2 - 2 + 2b + 1 - b^2;$$

$$\frac{B}{C} = \frac{1}{4}[2(1 - b) - a_2] - 1 + \frac{1 - b^2 + h}{2(1-b) - a_2}.$$

$$2(1 - b) - a_2 = 2(1 - b) - 2(\sigma - b) + \frac{2}{3}(1 - \sigma) - 4k\,\frac{\sqrt{\sigma} - 1}{\sqrt{\sigma}}$$

$$= \frac{8}{3}(1 - \sigma) + 4k\,\frac{1 - \sqrt{\sigma}}{\sqrt{\sigma}} = 4(1 - \sqrt{\sigma})\left[\frac{2}{3}(1 + \sqrt{\sigma}) + \frac{k}{\sqrt{\sigma}}\right].$$

$$1 - b^2 + h = 2(1-b^2) - 2(\sigma^2-b^2) - \frac{2}{9}(1-\sigma^2) - \frac{16}{3}k\,(1-\sqrt{\sigma}) - 4k^2\,\frac{1-\sigma}{\sigma}$$

$$= \frac{16}{9}(1-\sigma^2) - \frac{16}{3}k\,(1-\sqrt{\sigma}) - 4k^2\,\frac{1-\sigma}{\sigma}$$

$$= 4(1-\sqrt{\sigma})\left[\frac{4}{9}(1+\sigma)(1+\sqrt{\sigma}) - \frac{4}{3}k - k^2\,\frac{1+\sqrt{\sigma}}{\sigma}\right]$$

$\Rightarrow$

$$\frac{1-b^2+h}{2(1-b)-a_2} = \frac{\frac{4}{9}(1+\sigma)(1+\sqrt{\sigma}) - \frac{4}{3}k - k^2\,\frac{1+\sqrt{\sigma}}{\sigma}}{\frac{2}{3}(1+\sqrt{\sigma}) + \frac{k}{\sqrt{\sigma}}} = \frac{2}{3}(1+\sigma) - \frac{1+\sqrt{\sigma}}{\sqrt{\sigma}}\,k$$

$\Rightarrow$

$$\frac{B}{C} = (1-\sqrt{\sigma})\left[\frac{2}{3}\,(1+\sqrt{\sigma}) + \frac{k}{\sqrt{\sigma}}\right] - 1 + \frac{2}{3}(1+\sigma) - \frac{1+\sqrt{\sigma}}{\sqrt{\sigma}}k = \frac{1}{3} - 2k = \lambda.$$

Thus we see that the optimal choice (21) is in agreement with the equality case of the condition II $\leqq 0$ and the optimized Power-inequality (20) is sharp.

Theorem 1. The optimized Power-inequality (20) is sharp for the equality function (18). The two-parametric family of extremal functions f has the following initial coefficients:

$$(22)\quad \begin{cases} a_2 = 2(\sigma-b) - \frac{2}{3}(1-\sigma) + \frac{2}{3}(1-3\lambda)\left(1-\frac{1}{\sqrt{\sigma}}\right), \\ a_3 = a_2^2 + 1 - b^2 - 2(\sigma^2-b^2) - \frac{2}{9}(1-\sigma^2) - \frac{8}{9}(1-3\lambda)(1-\sqrt{\sigma}) + \frac{1}{9}(1-3\lambda)^2\left(1-\frac{1}{\sigma}\right), \\ \delta_2 = a_4 - 2a_2a_3 + \frac{13}{12}a_2^3 = \frac{2}{3}(1-b^2) - \frac{b}{2}a_2^2 - \lambda\left(a_3 - \frac{3}{4}a_2^2 + ba_2\right). \end{cases}$$

The parameters λ and σ are limited according to the two cases 1° and 2° of Figure 32.

This Theorem actually restates the result (24)/III.4.2, p. 285, of [1]. The parametrized coefficients (22), however, allow analyzing the equality cases more conveniently than the explicit formulae of III.4.2 of [1].

3. Integration of Löwner's Differential Equation

The generating function ϑ defined by (18)/VI.1.2 allows now determining the extremal f. This is found by integrating Löwner's differential equation in $S_R(b)$ which is given in [1], (11)/I.3.8, p. 68:

$$u\frac{\partial f}{\partial u}(1 - 2\cos\vartheta \cdot f + f^2) = f - f^3$$

$\Longleftrightarrow$

$$u\frac{\partial f}{\partial u}f^{-1}(f + f^{-1} - 2\cos\vartheta) = f^{-1} - f.$$

Multiply by the integrating factor (cf. [6])

$$\frac{3}{2}u^{1/2}(f^{1/2} + f^{-1/2}):$$

$$u\frac{\partial f}{\partial u}f^{-1}\cdot\frac{3}{2}u^{1/2}(f^{1/2}+f^{-1/2})(f+f^{-1}-2\cos\vartheta) + \frac{3}{2}u^{1/2}(f^{1/2}+f^{-1/2})(f-f^{-1}) = 0;$$

$$\frac{\partial f}{\partial u}f^{-1}\cdot\frac{3}{2}u^{3/2}[f^{3/2}+f^{-3/2}+(f^{1/2}+f^{-1/2})(1-2\cos\vartheta)] + \frac{3}{2}u^{1/2}(f^{3/2}-f^{-3/2}+f^{1/2}-f^{-1/2}) = 0$$

Denote, for brevity, $\cos\vartheta$ defined by (18)/VI.1.2

$$\text{(23)}\qquad \begin{cases} \cos\vartheta = \frac{1}{3} - \frac{K}{6}u^{-3/2}, & \sigma \leqq u \leqq 1, \\ K = 1 - 3\lambda = 6k. \end{cases}$$

Thus, $1 - 2\cos\vartheta = \frac{1}{3} + \frac{K}{3}u^{-3/2}$ and the equation assumes the form

$$\frac{\partial f}{\partial u}f^{-1}\cdot[\frac{3}{2}u^{3/2}(f^{3/2} + f^{-3/2}) + \frac{K + u^{3/2}}{2}(f^{1/2} + f^{-1/2})]$$

$$+ \frac{\partial}{\partial u}[u^{3/2}(f^{3/2} - f^{-3/2} + f^{1/2} - f^{-1/2})] = 0$$

$\Longleftrightarrow$

$$\frac{\partial f}{\partial u}[\frac{3}{2}u^{3/2}(f^{1/2} + f^{-5/2}) + \frac{K + u^{3/2}}{2}(f^{-1/2} + f^{-3/2})]$$

$$+ \frac{\partial}{\partial u}[u^{3/2}(f^{3/2} - f^{-3/2} + f^{1/2} - f^{-1/2})] = 0$$

$\Longleftrightarrow$

$$\frac{\partial f}{\partial u}\cdot\frac{\partial}{\partial f}[u^{3/2}(f^{3/2} - f^{-3/2}) + (K + u^{3/2})(f^{1/2} - f^{-1/2})]$$

$$+ \frac{\partial}{\partial u}[u^{3/2}(f^{3/2} - f^{-3/2} + f^{1/2} - f^{-1/2})] + \underbrace{\frac{\partial}{\partial u}[K(f^{1/2} - f^{-1/2})]}_{0} = 0$$

$\Longleftrightarrow$

$$(24)\quad \begin{cases} \dfrac{\partial f}{\partial u}\,\dfrac{\partial}{\partial f}[u,f] + \dfrac{\partial}{\partial u}[u,f] = 0; \\ [u,f] = u^{3/2}(f^{3/2} - f^{-3/2}) + (K + u^{3/2})(f^{1/2} - f^{-1/2}). \end{cases}$$

This result shows us that

$$2\,\frac{\partial}{\partial u}[u,f] = 0$$

which implies that

$$[u,f] = \text{constant with respect to } u.$$

Hence, the integrated equation assumes the form

$$[\sigma,f_\sigma] = [1,z]$$

i.e.

$$\begin{cases} \sigma^{3/2}(f_\sigma^{3/2}-f_\sigma^{-3/2}) + (K+\sigma^{3/2})(f_\sigma^{1/2}-f_\sigma^{-1/2}) = z^{3/2} - z^{-3/2} + (K+1)(z^{1/2}-z^{-1/2}); \\ f_\sigma = f(z,\sigma). \end{cases}$$

According to (18)/VI.1.2 the final function $f = f(z,b)$ is obtained by applying the left radial-slit-mapping to f_σ.

<u>Theorem 2</u>. The extremal function $f = f(z,b)$, determined by (18)/VI.1.2, is defined by the conditions

$$(25)\quad \sigma^{3/2}(f_\sigma^{3/2}-f_\sigma^{-3/2}) + (3\lambda-1+\sigma^{3/2})(f_\sigma^{1/2}-f_\sigma^{-1/2})$$

$$= z^{3/2} - z^{-3/2} + 3\lambda(z^{1/2}-z^{-1/2}),$$

$$(26)\quad b^{1/2}(f^{1/2} - f^{-1/2}) = \sigma^{1/2}(f_\sigma^{1/2} - f_\sigma^{-1/2}).$$

Here the parameters lie on the intervals

$$(27)\quad \begin{cases} \frac{1}{3} - \frac{4}{3}\sigma^{3/2} \leqq \lambda \leqq \frac{1}{3} + \frac{8}{3}\sigma^{3/2} \\ b \leqq \sigma \leqq 1. \end{cases}$$

The connection between the parameters (λ,σ) and the point (a_2,a_3) i the sharpness region I (cf. Figure 38) of the coefficient body is the one in (22)/VI.1.2.

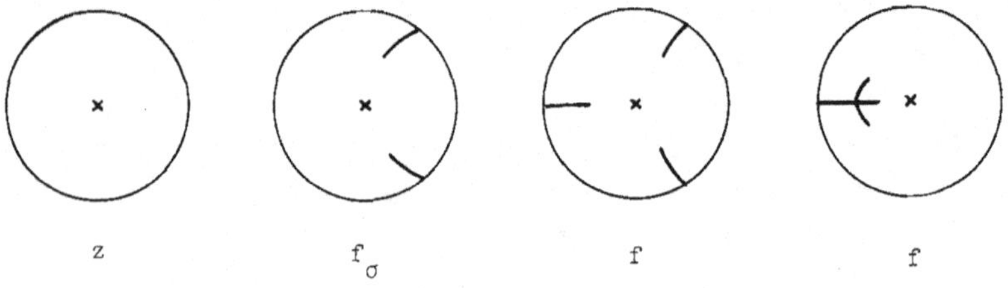

Figure 33.

2 § Schiffer's Differential Equation for Functions 2:3

1. The Class S(b)

Next, turn to the problem of generalizing the Power-inequality. We want to find out an inequality which is sharp for the $S_R(b)$-functions of the type 2:3. Thus, at least for the expected equality function we may assume that a condition similar to the Power-inequality holds. We will try to find out the proper condition by imitating the previous procedures. Here the development of the left side of Schiffer's equation plays a central role.

We start from Schiffer's differential equation connected with problems involving a_4 as the highest coefficient. Let us generalize the notations of IV.2.1, p. 294, of [1]. Thus, a necessary condition characterizing the boundary function f of the coefficient body (a_2,a_3,a_4) in $S(b)$ is

$$b^3\left(z\,\frac{f'}{f}\right)^2 \sum_{-3}^{3} c_\nu f^\nu = \sum_{-3}^{3} d_\nu z^\nu, \tag{1}$$

where

$$c_\nu = \bar{c}_{-\nu}, \quad d_\nu = \bar{d}_{-\nu}; \quad \nu = 0,1,2,3. \tag{2}$$

Consider especially the case where the solution of (1) is of the type 2:3. By using new parameters we rewrite (1) in the factorized form

$$\tau^3 b^3\left(z\frac{f'}{f}\right)^2 \frac{(f-f_1)^2(f-f_2)^2(f-f_3)(f-f_4)}{f^3} = \tau^3\,\frac{(z-z_1)^2(z-z_2)^2(z-z_3)^2}{z^3} \tag{3}$$

where $|\tau| = 1$. The geometrical meaning of the parameters f_ν, z_ν is illustrated by Figure 34.

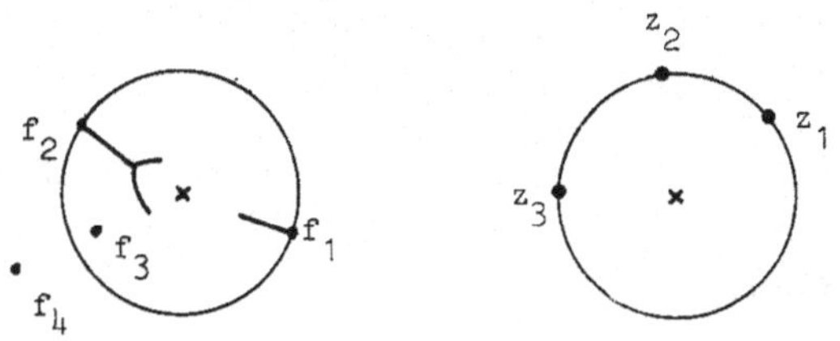

Figure 34.

Observe that in the case 2:2 the form of the left side is that of (3) but the right side is generalized (Figure 35):

$$(4) \qquad \tau^3 b^3 \left(z\frac{f'}{f}\right)^2 \frac{(f-f_1)^2(f-f_2)^2(f-f_3)(f-f_4)}{f^3} = \tau^3 \frac{(z-z_1)^2(z-z_2)^2(z-z_3)(z-z_4)}{z^3}$$

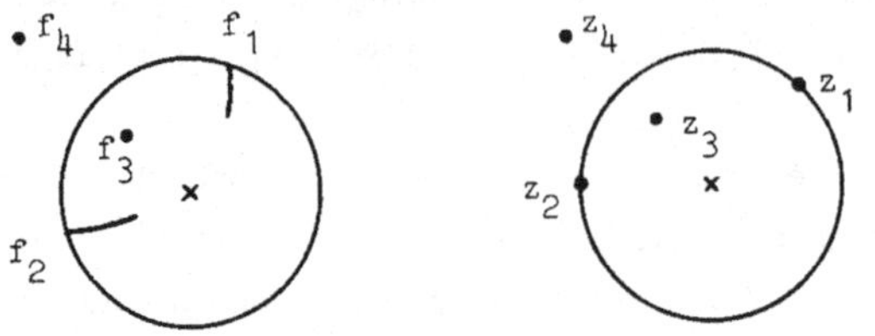

Figure 35.

The parameters f_ν and z_ν are not completely free. By comparing (1) and (4) we see that

$$\begin{cases} c_3 = \tau^3, \\ c_2 = -\tau^2[2(f_1 + f_2) + f_3 + f_4], \\ \dots \\ c_{-2} = -\tau^2[2f_1f_2f_3f_4(f_1 + f_2) + f_1^2f_2^2(f_3 + f_4)], \\ c_{-3} = \tau^3 f_1^2 f_2^2 f_3 f_4. \end{cases}$$

One symmetry condition involved in (2) is

$$c_3 = \bar{c}_{-3}$$

$$\Longleftrightarrow$$

$$\tau^3 f_1^2 f_2^2 f_3 f_4 = \tau^{-3}.$$

We may normalize this so that

$$(5) \qquad f_1 f_2 \sqrt{f_3 f_4} = -\tau^{-3}.$$

From this it follows further that

$$c_{-2} = \bar{c}_2$$

and that the other symmetry conditions (2) concerning the c_ν-numbers hold. Similarly, the symmetry conditions (2) for the d_ν-numbers are satisfied if in (4) we take (in (3) $z_3 = z_4$)

$$(6) \qquad z_1 z_2 \sqrt{z_3 z_4} = -\tau^{-3}.$$

We are interested in the integrated form of (3) which we write

$$(7) \quad g(f(z)) = \tau^{3/2} b^{3/2} \int \frac{(f-f_1)(f-f_2)\sqrt{(f-f_3)(f-f_4)}}{f^{5/2}} df$$

$$= y_{-3} z^{-3/2} + y_{-1} z^{-1/2} + y_1 z^{1/2} + y_3 z^{3/2} + \ldots$$

$$= \tau^{3/2} \int \frac{(z-z_1)(z-z_2)(z-z_3)}{z^{5/2}} dz.$$

From this we see that the development

$$\sum_{\nu=-2}^{\infty} y_{2\nu+1} z^{\frac{2\nu+1}{2}}$$

is finite in the case 2:3 in question. This involves an important property of the extremal function f which can be utilized when looking for the generalized Power-inequality for f. Observe, again, that (4) does not guarantee this finiteness property for the elliptic extremal functions 2:2.

In order to find the y_ν-coefficients it is useful to form first the development of the integrand. Introduce the following notations together with the choice of the branch of the square root:

$$\begin{cases} \sqrt{f_3 f_4 - (f_3 + f_4) f + f^2} = -\sqrt{f_3 f_4}(1 - pf + qf^2)^{1/2}; \\ p = \bar{f}_3 + \bar{f}_4, \quad q = \bar{f}_3 \bar{f}_4; \end{cases}$$

$$\begin{cases} (f - f_1)(f - f_2) = f_1 f_2 (1 - rf + sf^2); \\ r = \bar{f}_1 + \bar{f}_2, \quad s = \bar{f}_1 \bar{f}_2, \quad D = \frac{p^2}{4} - q; \end{cases}$$

$$(f - f_1)(f - f_2)\sqrt{f_3 f_4 - (f_3 + f_4)f + f^2}$$

$$= \underbrace{-f_1 f_2 \sqrt{f_3 f_4}}_{\tau^{-3}} \left[1 - (\frac{p}{2} + r)f + (s + \frac{pr}{2} - \frac{D}{2})f^2 + \frac{1}{2}(rD - ps - \frac{pD}{2})f^3 + \ldots \right]$$

$\Rightarrow$

$$g(f) = \tau^{-3/2} b^{3/2} \int \left[f^{-5/2} - (\frac{p}{2} + r)f^{-3/2} + (s + \frac{pr}{2} - \frac{D}{2})f^{-1/2} + \frac{1}{2}(rD - ps - \frac{pD}{2})f^{1/2} + \ldots \right] df$$

$$= \tau^{-3/2} b^{3/2} \left[-\frac{2}{3} f^{-3/2} + (p+2r)f^{-1/2} + (2s+pr-D)f^{1/2} + \frac{1}{3}(rD - ps - \frac{pD}{2})f^{3/2} + \ldots \right].$$

In the number $\zeta = z^{1/2}$ we have

$$f^{-3/2} = b^{-3/2}\zeta^{-3} - \frac{3}{2} b^{-3/2} a_2 \zeta^{-1} + b^{-3/2}(\frac{15}{8} a_2^2 - \frac{3}{2} a_3)\zeta - b^{-3/2}(\frac{3}{2} a_4 - \frac{15}{4} a_2 a_3 + \frac{35}{16} a_2^3)\zeta^3 + .$$

$$f^{-1/2} = b^{-1/2}\zeta^{-1} - \frac{1}{2} b^{-1/2} a_2 \zeta + b^{-1/2}(\frac{3}{8} a_2^2 - \frac{1}{2} a_3)\zeta^3 + \ldots,$$

$$f^{1/2} = b^{1/2}\zeta + \frac{1}{2} b^{1/2} a_2 \zeta^2 + \ldots,$$

$$f^{3/2} = b^{3/2}\zeta^3 + \ldots;$$

$$g(f(z)) = y_{-3}\zeta^{-3} + y_{-1}\zeta^{-1} + y_1\zeta + y_3\zeta^3 + \ldots .$$

The y_ν-coefficients are expressed, as before, by using the abbreviations

$$u_1 = -y_{-1}, \quad u_3 = -3y_{-3};$$

$$(8)\quad \begin{cases} y_{-3} = -\frac{2}{3}\,\tau^{-3/2} = -\frac{u_3}{3}, \\ y_{-1} = \tau^{-3/2}[a_2 + (p+2r)b] = -u_1, \\ y_1 = \tau^{-3/2}[a_3 - \frac{5}{4}\,a_2^2 - \frac{b}{2}(p+2r)a_2 + b^2(2s+pr-D)], \\ y_3 = \tau^{-3/2}[a_4 - \frac{5}{2}\,a_2a_3 + \frac{35}{24}\,a_2^3 - \frac{b}{2}(p+2r)(a_3 - \frac{3}{4}a_2^2) + \frac{1}{2}b^2(2s+pr-D)a_2 \\ \qquad + \frac{1}{3}b^3(rD - ps - \frac{1}{2}pD)]. \end{cases}$$

The condition expected to be sharp in $S(b)$ with the equality-function of the type 2:3 is obtained from these y_ν-numbers by imitating the truncated Power-inequality ((9)/III.1.5, p. 180, [1]):

$$(9)\qquad \operatorname{Re}\sum_1^3 u_k y_k \leq \sum_1^3 \frac{|u_k|^2}{k} \qquad (u_2 = 0).$$

2. The Class $S_R(b)$.

Let us apply the previous formulae in the special case $S_R(b)$ where the notations f_ν mean the following numbers, illustrated in Figure 36:

$$f_1 = 1, \quad f_2 = -1, \quad f_3 = \rho \in (0,1], \quad f_4 = \rho^{-1}.$$

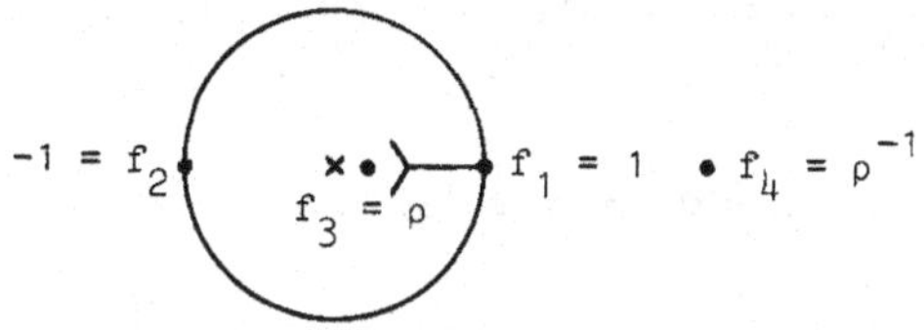

Figure 36.

Further, we have in $S_R(b)$:

$$\begin{cases} \tau = 1, \\ p = \rho + \rho^{-1} = R \geqq 2, \\ q = 1, \\ D = \frac{R^2}{4} - 1, \\ r = 0, \\ s = -1; \end{cases}$$

$$\begin{cases} p + 2r = R, \\ 2s + pr - D = -\frac{R^2}{4} - 1, \\ rD - ps - \frac{1}{2}\,pD = \frac{3}{2}\,R - \frac{1}{8}\,R^3. \end{cases}$$

Thus, the formulae (8) yield

$$(10)\quad \begin{cases} u_3 = 2, \\ u_1 = -y_{-1} = -(a_2 + Rb), \\ y_1 = a_3 - \frac{5}{4}\,a_2^2 - \frac{1}{2}\cdot \underbrace{bR}_{-a_2-u_1} \cdot a_2 - b^2(\frac{R^2}{4} + 1) = \delta_1 + \frac{1}{2}\,u_1 a_2 - (\frac{R^2}{4} + 1)b^2, \\ y_3 = a_4 - \frac{5}{2}\,a_2 a_3 + \frac{35}{24}\,a_2^3 - \frac{1}{2}\cdot \underbrace{bR}_{-a_2-u_1} \cdot (a_3 - \frac{3}{4}\,a_2^2) - \frac{b^2}{2}(\frac{R^2}{4} + 1)a_2 + \frac{b^3}{3}(\frac{3}{2}R - \frac{R^3}{8} \\ \quad = \delta_2 + \frac{1}{2}\,\delta_1 u_1 - \frac{b^2}{2}(\frac{R^2}{4} + 1)a_2 + \frac{b^3}{3}(\frac{3}{2}R - \frac{R^3}{8}). \end{cases}$$

Here δ_1 and δ_2 are the combinations (3)/VI.1.1. We can still shorten the notation by introducing

$$(11)\quad \begin{cases} d_1 = a_3 - a_2^2 - b^2 = \delta_1 - \frac{a_2^2}{4} - b^2, \\ d_2 = a_4 - 2a_2 a_3 + a_2^3 - b^2 a_2 = \delta_2 - \frac{a_2^3}{12} - b^2 a_2. \end{cases}$$

In d_1 and d_2 the numbers y_1 and y_3 assume the form

$$(12)\qquad \begin{cases} y_1 = d_1 - \frac{u_1^2}{4}, \\ y_3 = d_2 + \frac{1}{2}\, u_1 d_1 + \frac{u_1^3}{24}. \end{cases}$$

By aid of (12) we can write the inequality (9) in the form

$$(13)\qquad \frac{1}{2}(u_1 y_1 + u_3 y_3 - u_1^2 - \frac{u_3^2}{3}) = d_2 + u_1 d_1 - \frac{2}{3} - \frac{u_1^2}{2} - \frac{u_1^3}{12} \leqq 0.$$

This condition is expected to hold in $S_R(b)$. In order to verify this we must, again, to find out proper Löwner-identities.

3 § A Generalized Power-inequality for $S_R(b)$-functions

1. Löwner-identities

Let us start from the formulae (1)/VI.1.1:

$$\begin{cases} a_2 = -2\displaystyle\int_b^1 \cos\vartheta\, du, \\ a_3 - a_2^2 = -2\displaystyle\int_b^1 u\cos 2\vartheta\, du = b^2 - 1 + 4\int_b^1 u\sin^2\vartheta\, du. \end{cases}$$

From (2)/VI.1.1 we see that

$$d_2 + b^2 a_2 = a_4 - 2a_2a_3 + a_2^3 = -4\int_b^1 u \underbrace{\cos 2\vartheta}_{1-2\sin^2\vartheta} \left(\int_1^u \cos\vartheta\, d\tilde{u}\right) - 2\int_b^1 u^2\cos 3\vartheta\, du.$$

$$= -4\int_b^1 u\left(\int_1^u \cos\vartheta\, d\tilde{u}\right)du + 8\int_b^1 u\sin^2\vartheta\left(\int_1^u \cos\vartheta\, d\tilde{u}\right)du - 2\int_b^1 u^2\cos 3\vartheta\, du$$

Here is

$$\int_b^1 u\left(\int_1^u \cos\vartheta\, d\tilde{u}\right)du = \int_b^1\left(\int_1^u \cos\vartheta\, d\tilde{u}\right)d\left(\frac{u^2}{2}\right)$$

$$= \Big/_b^1 \frac{u^2}{2}\int_1^u \cos\vartheta\, d\tilde{u} - \frac{1}{2}\int_b^1 u^2\cos\vartheta\, du = \frac{b^2}{2}\int_b^1 \cos\vartheta\, du - \frac{1}{2}\int_b^1 u^2\cos\vartheta\, du$$

$\Rightarrow$

$$-4\int_b^1 u\left(\int_1^u \cos\vartheta\, d\tilde{u}\right)du = \underbrace{-2b^2\int_b^1 \cos\vartheta\, du}_{b^2 a_2} + 2\int_b^1 u^2\cos\vartheta\, du.$$

Thus

(2) $$d_2 = a_4 - 2a_2a_3 + a_2^3 - b^2 a_2$$

$$= 2\int_b^1 u^2 \cos\vartheta\, du + 8\int_b^1 u \sin^2\vartheta \left(\int_1^u \cos\vartheta\, d\tilde{u}\right) du - 2\int_b^1 u^2 \cos 3\vartheta\, du.$$

Because

$$\cos\vartheta - \cos 3\vartheta = 4\cos\vartheta \sin^2\vartheta$$

we have, according to (1) and (2),

$$(3)\quad \begin{cases} d_2 = 8\int_b^1 u^2 \cos\vartheta \sin^2\vartheta\, du + 8\int_b^1 u \sin^2\vartheta \left(\int_1^u \cos\vartheta\, d\tilde{u}\right) du, \\ d_1 = -1 + 4\int_b^1 u \sin^2\vartheta\, du. \end{cases}$$

This gives for the combination $d_2 + u_1 d_1$ in (13)/VI.2.2 the Löwner-presentation

$$(4)\quad d_2 + u_1 d_1 = -u_1 + 8\int_b^1 u \sin^2\vartheta \left(u\cos\vartheta - \int_u^1 \cos\vartheta\, d\tilde{u} + \frac{u_1}{2}\right) du.$$

2. An Inequality of Power-type Expected to be Sharp in $S_R(b)$ with the Equality Function 2:3.

The relatively simple Löwner-expression (4) allows rewriting (13) as follows:

$$d_2 + u_1 d_1 = -u_1 + 8\int_b^1 u \sin^2\vartheta \left(u\cos\vartheta - \int_u^1 \cos\vartheta\, d\tilde{u} + \frac{u_1}{2}\right) du \leqq \frac{2}{3} + \frac{u_1^2}{2} + \frac{u_1^3}{12}$$

$\Longleftrightarrow$

$$\int_b^1 u \sin^2\vartheta \left(u\cos\vartheta - \int_u^1 \cos\vartheta\, d\tilde{u} + \lambda\right) du = \frac{(\lambda + 1)^3}{12}; \quad \lambda = \frac{u_1}{2}.$$

Adopt still an abbreviation H:

$$(5)\qquad \begin{cases} \int_b^1 u \sin^2 \vartheta \, H(u)du \leqq \dfrac{(\lambda + 1)^3}{12}, \\ H(u) = u \cos \vartheta - \int_u^1 \cos \vartheta \, d\tilde{u} + \lambda, \\ \lambda = \dfrac{u_1}{2}. \end{cases}$$

This is the inequality we expect to be sharp in $S_R(b)$ and such that the equality function is of the type 2:3. We will briefly return to the validity question later on. At first, let us consider the equality.

As in VI.1.2 there are two factors in the integrand which determine the value of the integral in question. Here, however, maximalization does not follow from that of the integrand. The fact that the equality is reached can be checked by remembering that (5) was constructed to be sharp for the 2:3-cases. This implies that we have to define $\cos \vartheta$ for the equality function as follows (Figure 37):

$$(6)\qquad \cos \vartheta = \begin{cases} -1 & , \; b \leqq u \leqq \sigma_1, \\ 1 & , \; \sigma_1 < u \leqq \sigma_2, \\ \dfrac{1}{3} + \dfrac{k}{3} u^{-3/2}, & \sigma_2 \leqq u \leqq 1; \quad k = \dfrac{1 - 3\lambda}{2}. \end{cases}$$

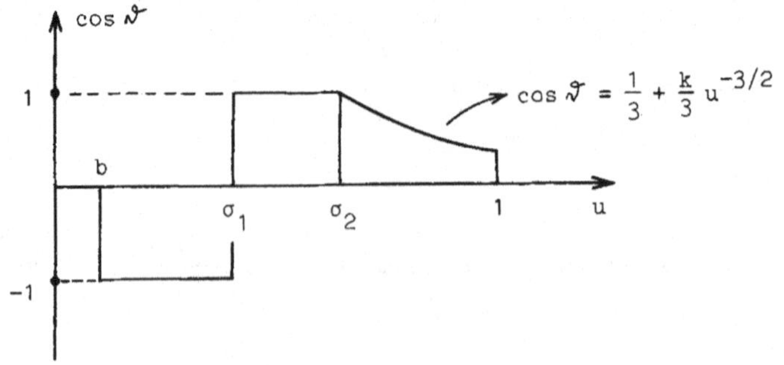

Figure 37.

The numbers σ_2 and λ are connected so that

$$(7) \qquad 1 = \frac{1}{3} + \frac{k}{3}\sigma_2^{-3/2} \Rightarrow \sigma_2 = \left(\frac{k}{2}\right)^{2/3} = \left(\frac{1-3\lambda}{4}\right)^{2/3} \in [b,1]$$

$\Rightarrow$

$$-1 \leqq \lambda \leqq \frac{1}{3} - \frac{4}{3} b^{3/2}.$$

The form obtained for $\cos\vartheta$ in (6) is based on the knowledge of the structure that holds for solutions of Schiffer's differential equation.

When determining the value of the left side of (5) we may utilize the connection between the H- and K-functions ((9)/VI.1.1):

$$(8) \qquad \begin{cases} H(u) = -K(u) + u(1 - \cos\vartheta), \\ K(u) = \int_u^1 \cos\vartheta \, d\tilde{u} - u(2\cos\vartheta - 1) - \lambda; \\ K(u) = 0 \quad \text{for} \quad \cos\vartheta = \frac{1}{3} + \frac{k}{3} u^{-3/2}. \end{cases}$$

For $\cos\vartheta = \frac{1}{3} + \frac{k}{3} u^{-3/2}$ is thus

$$\begin{cases} H(u) = \frac{2}{3}u - \frac{k}{3} u^{-1/2}, \\ \sin^2\vartheta = \dfrac{8 - k^2 u^{-3} - 2ku^{-3/2}}{9}; \end{cases}$$

$$27\int_b^1 u \sin^2\vartheta \, H(u)du = 27\int_{\sigma_2}^1 u\sin^2\vartheta \, H(u)du = \int_{\sigma_2}^1 (16u^2 - 12ku^{1/2} + k^3u^{-5/2})du$$

$$= \frac{16}{3}(2 - \underbrace{\sigma_2^3}_{(\frac{k}{2})^2}) - 8k + 4k^2 + \frac{4}{3}k^2 - \frac{2}{3}k^3$$

$$= \frac{2}{3}(2-k)^3 = \frac{2}{3}\cdot\frac{27}{8}(\lambda + 1)^3$$

$\Rightarrow$

$$\int_b^1 u \sin^2 \vartheta \, H(u)du = \frac{(\lambda+1)^3}{12}.$$

Next, determine the coefficients a_2 and a_3 for the equality function (6).

$$-\frac{1}{2} a_2 = \int_b^{\sigma_1} (-1)du + \int_{\sigma_1}^{\sigma_2} du + \int_{\sigma_2}^{1} \left(\frac{1}{3} + ku^{-3/2}\right)du$$

$$= \frac{1}{3} - \frac{2}{3}k + b - 2\sigma_1 + \frac{2}{3}\sigma_2 + \frac{2}{3}\underbrace{k\sigma_2^{-1/2}}_{2\sigma_2}$$

$$= \frac{1}{3} + b - \frac{4}{3}\sigma_2^{3/2} - 2\sigma_1 + 2\sigma_2.$$

$$-\frac{1}{2}(a_3 - a_2^2) = \int_b^1 u(2\cos^2\vartheta - 1)du = \frac{1}{2}(b^2 - 1) + 2\int_b^1 u \cos^2 \vartheta \, du$$

⇒

$$-\frac{1}{4}(a_3 - a_2^2) + \frac{1}{4}(1 - b^2) = \int_b^1 u \cos^2 \vartheta \, du = \int_b^{\sigma_2} udu + \int_{\sigma_2}^1 u \, \frac{1 + 2ku^{-3/2} + k^2u^{-3}}{9} \, d$$

⇒

$$9[-\frac{1}{4}(a_3 - a_2^2) + \frac{1}{4}(1 - b^2) + \frac{1}{2}(b^2 - \sigma_2^2)] = \int_{\sigma_2}^1 (u + 2ku^{-1/2} + k^2u^{-2})du$$

$$= \frac{1}{2} + \underbrace{4k}_{8\sigma_2^{3/2}} - \underbrace{k^2}_{4\sigma_2^3} - \left(\frac{\sigma_2^2}{2} + \underbrace{4k\sigma_2^{1/2}}_{8\sigma_2^2} - \underbrace{k^2\sigma_2^{-1}}_{4\sigma_2^2}\right) = \frac{1}{2} + 8\sigma_2^{3/2} - \frac{9}{2}\sigma_2^2 - 4\sigma_2^3.$$

Collect the results for the equality function.

Theorem 1. The equality function f of (5), defined by (6), has the coefficients

$$(9) \quad \begin{cases} a_2 = -\frac{2}{3} - 2b + 4\sigma_1 - 4\sigma_2 + \frac{8}{3}\sigma_2^{3/2}, \\ a_3 = a_2^2 + \frac{7}{9} + b^2 - \frac{32}{9}\sigma_2^{3/2} + \frac{16}{9}\sigma_2^3, \end{cases}$$

where the two parameters σ_1 and σ_2 are restricted so that (Figure 37, conditions (6) and (7)):

$$(10)\quad b \leqq \sigma_1 \leqq \sigma_2 \leqq 1; \quad \sigma_2 = \left(\frac{1-3\lambda}{4}\right)^{2/3}.$$

The limitation $\sigma_2 \in [b,1]$ implies for λ

$$(11)\quad -1 \leqq \lambda \leqq \frac{1}{3} - \frac{4}{3}\, b^{3/2}.$$

3. The Boundary Curves of the Sharpness Region

Theorem 1/VI.1.2 and Theorem 1/VI.3.2 give parametric presentation for points (a_2,a_3) which lie in the region where the corresponding inequalities (20)/VI.1.2 and (5)/VI.3.2, concerning the coefficient body (a_2,a_3,a_4), are sharp. We are interested especially in the boundary curves of the sharpness domains. In Figure 38 there are introduced the notations 1, 2, 3, 1', 2', 3' for the curves which form the boundary or separate different types of equality functions. - The meaning of the arcs 1°, 2° and 3° will be clarified later on.

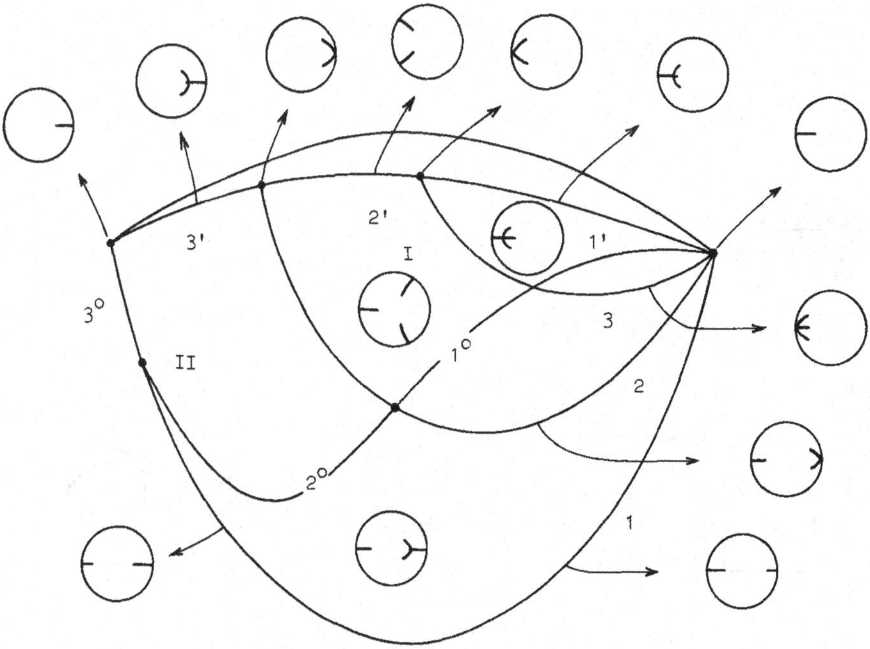

Figure 38.

In the following there is a list of parametric presentations of the curves 1,...,3'.

Curve 1.

In (9) take

$$\begin{cases} \sigma_2 = 1; \quad k = 2; \quad \lambda = -1; \\ \sigma_1 \in [b,1]. \end{cases}$$

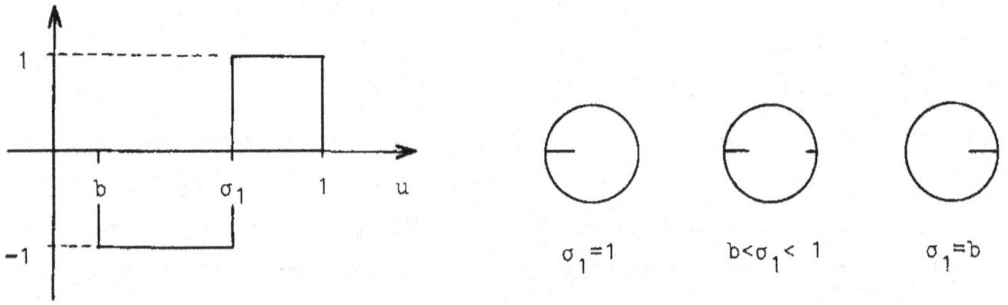

Figure 39.

$$1: \begin{cases} a_2 = -2 - 2b + 4\sigma_1; \\ a_3 = a_2^2 + b^2 - 1. \end{cases}$$

Curve 2.

In (9) take

$$\sigma_1 = \sigma_2 = \left(\frac{1-3\lambda}{4}\right)^{3/2} \in [b,1].$$

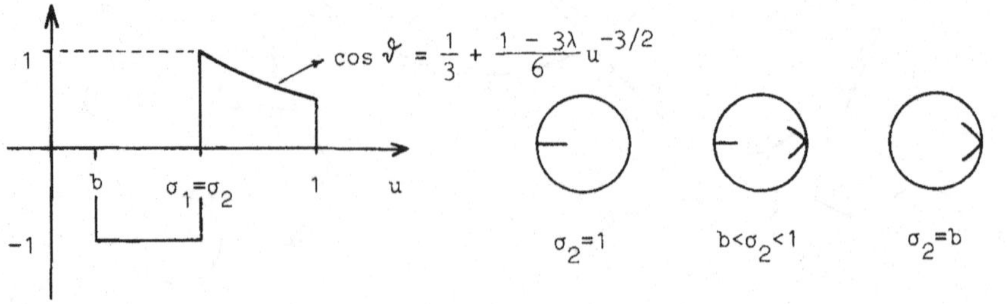

Figure 40.

$$2: \begin{cases} a_2 = -\frac{2}{3} - 2b + \frac{8}{3}\sigma_2^{3/2}, \\ a_3 = a_2^2 + \frac{7}{9} + b^2 - \frac{32}{9}\sigma_2^{3/2} + \frac{16}{9}\sigma_2^3 = a_2^2 + b^2 - 1 + \frac{16}{9}(1 - \sigma_2^{3/2})^2; \end{cases}$$

$$2: \quad a_3 = a_2^2 + b^2 - 1 + \frac{1}{4}[2(1 - b) - a_2]^2.$$

Curve 3.

This curve, which separates the regions of 1:3 and 3:3, is determined in III.3.2 of [1]:

$$3: \quad a_3 = \frac{5}{4}a_2^2 - (1 + 3b)a_2 + 6b(1 - b).$$

Curve 1'.

In (22)/VI.1.2 take

$$\lambda = \frac{1}{3} + \frac{8}{3}\sigma^{3/2}, \quad \sigma \in [b,1].$$

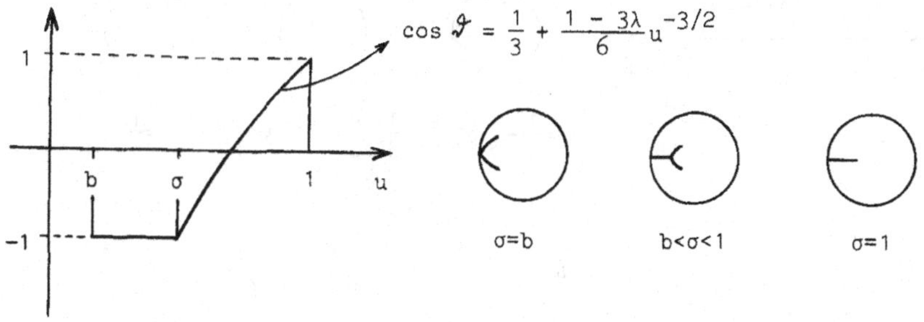

Figure 41.

$$1': \begin{cases} a_2 = -\frac{2}{3} - 2b + 8\sigma - \frac{16}{3}\sigma^{3/2}, \\ a_3 = a_2^2 + \frac{7}{9} + b^2 - \frac{144}{9}\sigma^2 + \frac{64}{9}\sigma^{3/2} + \frac{64}{9}\sigma^3. \end{cases}$$

Curve 2'.

In (22)/VI.1.2 take

$$\sigma = b; \quad \frac{1}{3} \le \lambda \le \frac{1}{3} + \frac{8}{3} b^{3/2}.$$

$$\cos \vartheta = \frac{1}{3} + \frac{1 - 3\lambda}{6} u^{-3/2}$$

$$\lambda = \frac{1}{3} \qquad \frac{1}{3} < \lambda < \frac{1}{3} + \frac{8}{3} b^{3/2} \qquad \lambda = \frac{1}{3} + \frac{8}{3} b^{3/2}$$

$$\cos \vartheta = \frac{1}{3} + \frac{1 - 3\lambda}{6} u^{-3/2}$$

$$\lambda = \frac{1}{3} - \frac{4}{3} b^{3/2} \qquad \frac{1}{3} - \frac{4}{3} b^{3/2} < \lambda < \frac{1}{3} \qquad \lambda = \frac{1}{3}$$

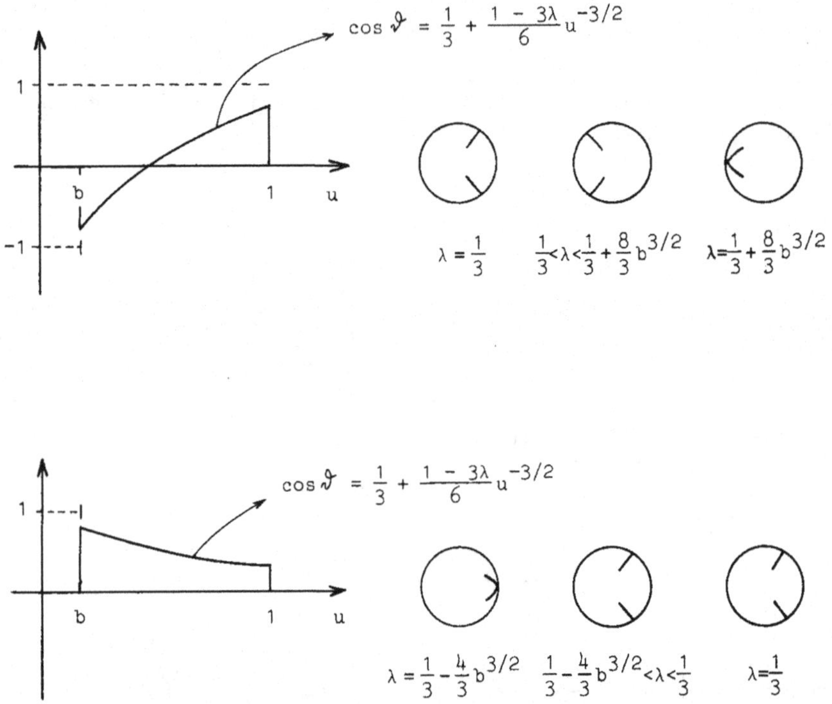

Figure 42.

$$2': \begin{cases} a_2 = -\frac{2}{3}(1-b) + \frac{2}{3}(1-3\lambda)(1-b^{-1/2}), \\ \\ a_3 = a_2^2 + \frac{7}{9}(1-b^2) - \frac{8}{9}(1-3\lambda)(1-b^{1/2}) + \frac{1}{9}(1-3\lambda)^2(1-b^{-1}). \end{cases}$$

Curve 3'.

In (9) take

$$\begin{cases} \sigma_1 = b, \quad \sigma_2 \in [b,1] \\ b \leqq \sigma_2 \leqq 1 \Leftrightarrow -1 \leqq \lambda \leqq \frac{1}{3} - \frac{4}{3} b^{3/2}. \end{cases}$$

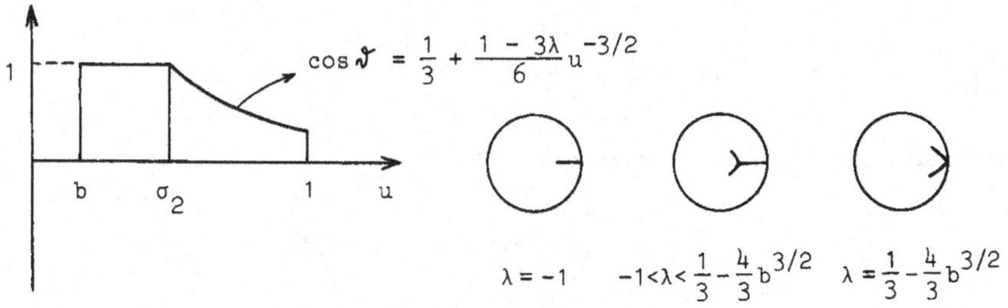

Figure 43.

$$3': \begin{cases} a_2 = -\frac{2}{3} + 2b - 4\sigma_2 + \frac{8}{3}\sigma_2^{3/2}, \\ a_3 = a_2^2 + \frac{7}{9} + b^2 - \frac{32}{9}\sigma_2^{3/2} + \frac{16}{9}\sigma_2^3. \end{cases}$$

Let us denote by I the region bounded by the curves 1', 2' and 2. According to the results of III.2.5, p. 213, of [1] we know that the optimized Power-inequality (20)/VI.1.2 is sharp in I and the upper bound of a_4 ((4)/ III.2.5, p. 211, [1]) is further maximized in a_3 on the parabola

$$1^\circ: \quad a_3 = -\frac{1}{4} a_2^2 + (2-3b)a_2.$$

The curves 1° and 2 meet each other at the points

$$(12) \qquad \begin{cases} a_2 = 2 - 2b, \\ a_3 = 3 - 8b + 5b^2; \end{cases} \qquad \begin{cases} a_2 = -\frac{2}{3} b, \\ a_3 = -\frac{4}{3} b + \frac{17}{9} b^2. \end{cases}$$

The curve 1° was used in [1] for maximizing the coefficient a_4 in $S_R(b)$ for two intervals of b.

In order to extend the result beyond these intervals we introduce the region II bounded by the curves 1, 2 and 3' (Figure 38). The equality of the condition (5)/VI.3.1 is reached in this region II. This is seen by using the coefficients a_2 and a_3 of the equality function, given by (9)/VI.3.2. Instead of σ_2 it is advisable to use the parameter r defined so that

$$\sigma_2 = \sigma_1 + r(1-\sigma_1), \quad 0 \leqq r \leqq 1. \tag{13}$$

The conditions (9)/VI.3.2 thus assume the form

$$\begin{cases} a_2 = -\frac{2}{3} - 2b + 4r(\sigma_1 - 1) + \frac{8}{3}[\sigma_1 + r(1-\sigma_1)]^{3/2}, \\ a_3 = a_2^2 + \frac{7}{9} + b^2 - \frac{32}{9}[\sigma_1 + r(1-\sigma_1)]^{3/2} + \frac{16}{9}[\sigma_1 + r(1-\sigma_1)]^3; \\ 0 \leqq r \leqq 1, \quad b \leqq \sigma_1 \leqq 1. \end{cases} \tag{14}$$

The square $[0,1] \times [b,1] \ni (r,\sigma_1)$ is mapped by (14) onto the domain II, as can be checked by aid of computers. Thus II is confirmed to be the sharpness region of the condition (5)/VI.3.1.

4 § Consequences for the Second Coefficient Body in $S_R(b)$

1. Application to the Second Coefficient Body

Let us express the inequality (5)/VI.3.2 by using the abbreviations d_1 and d_2 introduced in (11)/VI.3.1:

$$(1)\qquad \begin{cases} d_1 = a_3 - a_2^2 - b^2, \\ d_2 = a_4 - 2a_2a_3 + a_2^3 - b^2a_2; \\ d_2 + 2\lambda(d_1 + 1) = 8\int_b^1 u\sin^2\vartheta\, H(u)du \leqq \frac{2}{3}(\lambda + 1)^3. \end{cases}$$

This inequality is proved by O. Jokinen for

$$(2)\qquad -1 \leqq \lambda \leqq 0.$$

The proof is based on global alternations of the generating function ϑ by aid of proper permutations. The rather cumbersome proof, given in [6], is omitted here.

By aid of numerical checking the validity of (1) seems to be extendable to the interval

$$(3)\qquad -1 \leqq \lambda \leqq \frac{1}{3} - \frac{4}{3}b^{3/2}.$$

For brevity, introduce the notation

$$(4)\qquad \lambda + 1 = x$$

and rewrite (1) in the form

$$(5)\qquad a_4 - 2a_2a_3 + a_2^3 - b^2a_2 + 2(x - 1)(a_3 - a_2^2 - b^2 + 1) - \frac{2}{3}x^3 \leqq 0.$$

Now optimize this inequality by choosing x so that the derivative of the left side (5) vanishes. This requires that

$$(6)\qquad x = x_o = \sqrt{a_3 - a_2^2 - b^2 + 1} \leqq 1.$$

This number exists in that part of the coefficient body which lies below the parabola $a_3 = a_2^2 + b^2$ and above the lower boundary curve 1 (Figure 38) i.e.

$$a_2^2 + b^2 - 1 \leqq a_3 \leqq a_2^2 + b^2.$$

The optimized inequality, expressed by using the abbreviation x_o, assumes the form

$$(7) \qquad \begin{cases} a_4 \leqq 2a_2a_3 - a_2^3 + b^2a_2 - 2(x_o - 1)(a_3 - a_2^2 - b^2 + 1) + \frac{2}{3}x_o^3 = F, \\ x_o = \sqrt{a_3 - a_2^2 - b^2 + 1}. \end{cases}$$

The coefficients a_2 and a_3 of the equality function of (7) are determined by the conditions (9), (10)/VI.3.2. As was pointed out in VI.3.3 reaching the equality of (7) in the whole region II requires that the parameter λ traces the entire interval (3). If only the interval (2) is used then for certain values of b there is left a gap in II, above the parabola $a_3 = a_2^2 + b^2$, where the validity of (7) remains open. Actually, if b is limited so that

$$\frac{1}{3} - \frac{4}{3}b^{3/2} \leqq 0$$

i.e.

$$(8) \qquad 0.396\cdot850 = (\tfrac{1}{4})^{2/3} \leqq b < 1$$

then for λ the interval (2) is available.

Theorem 1. The optimized inequality (7) extends the sharp estimation of a_4 in terms of b, a_2 and a_3 to the region II belonging to the algebraic part of the coefficient body (a_2, a_3, a_4). The estimation can be proved to be sharp on the whole of II for the values $(\frac{1}{4})^{2/3} \leqq b < 1$ if the interval (2) is used. If the interval (3) is available then the sharp estimation can be shown to hold in the whole of II for $0 < b < 1$.

2. Maximizing a_4 in b and a_2.

The optimized inequality (7) allows further reducing the arguments in which a_4 can be sharply estimated. Suppose first that b and a_2 are fixed and choose a_3 so that the upper bound F of a_4 in (7) is maximized. For brevity, we may use the variable x_o instead of a_3 in F for which we thus have

$$\begin{cases} F(x_o) = a_2^3 + (3b^2 - 2)a_2 + 2(a_2 + 1)x_o^2 - \frac{4}{3}x_o^3; \\ F'(x_o) = 4(a_2 + 1)x_o - 4x_o^2. \end{cases}$$

The roots of $F'(x_o) = 0$ are thus $x_o = 0$ and $x_o = a_2 + 1$. Because F is of third degree in x_o, we see immediately that (cf. Figure 44):

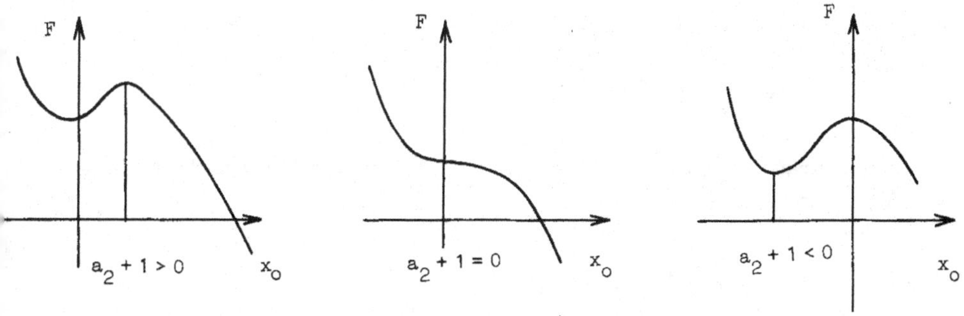

Figure 44.

$$(9) \quad \max F(x_o) = \begin{cases} F(a_2 + 1) = a_2^3 + (3b^2 - 2)a_2 + \frac{2}{3}(a_2 + 1)^3 & \text{for } a_2 + 1 \geqq 0, \\ F(0) = a_2^3 + (3b^2 - 2)a_2 & \text{for } a_2 + 1 \leqq 0. \end{cases}$$

The choice of x and consequently that of $\lambda = x - 1$ belonging to the maximum is

$$x_o = a_2 + 1 \Rightarrow \lambda = a_2,$$

$$x_o = 0 \qquad \Rightarrow \lambda = -1.$$

Thus, if the use of (7) and hence that of (1) is limited to the values

$$(10) \qquad a_2 \leqq 0$$

we obtain for a_4 an upper bound which is valid for all the values $a_2 \leqq 0$.

The point (a_2,a_3) in which the maximum (9) is reached satisfies the conditions

$$x_o = \sqrt{a_3 - a_2^2 - b^2 + 1} = a_2 + 1 \quad \text{or} \quad 0.$$

The latter of these gives an arc of the lower boundary curve 1 of (a_2,a_3). Denote this arc by 3° (Figure 38):

$$3^o: \qquad a_3 = a_2^2 - 1 + b^2.$$

The former defines the parabola (Figure 38):

$$2^o: \qquad a_3 = 2a_2^2 + 2a_2 + b^2.$$

The parabola on which a_4 was maximized in I for given numbers b and a_2 was found above to be

$$1^o: \qquad a_3 = -\frac{1}{4}a_2^2 + (2 - 3b)a_2.$$

The parabolas 2° and 1° meet, having a joint tangent, at the latter point (12)/VI.3.3

$$(11) \qquad a_2 = -\frac{2}{3}b, \quad a_3 = -\frac{4}{3}b + \frac{17}{9}b^2.$$

This is on the joint border arc 2 of I and II. The parabolas 3° and 2° have also a joint tangent at the point

$$(12) \qquad a_2 = -1, \quad a_3 = b^2.$$

The maxima found and the location of the arcs 1°, 2° and 3° with respect to the points (11) and (12) now suggest the use of (9) up to the point $a_2 = -\frac{2}{3}b < 0$. Thus we are sure that all the estimates in (13) are valid in their corresponding intervals:

Theorem 2. The numbers

$$
(13)\quad \begin{cases}
a_4(b,a_2;1^\circ) = -\frac{7}{12}a_2^3 + (2-\frac{9}{2}b)a_2^2 + \frac{2}{3}(1-b^3), & -\frac{2}{3}b \leqq a_2 \leqq 2(1-b), \\
a_4(b,a_2;2^\circ) = a_2^3 + (3b^2-2)a_2 + \frac{2}{3}(a_2+1)^3, & -1 \leqq a_2 \leqq -\frac{2}{3}b, \\
a_4(b,a_2;3^\circ) = a_2^3 + (3b^2-2)a_2, & -2(1-b) \leqq a_2 \leqq -1 \quad (b \leqq \frac{1}{2}).
\end{cases}
$$

define an upper bound of a_4 for given values of a_2 and b in $S_R(b)$. This upper bound is sharp so far as the arcs 1° and 2° lie in I and II, respectively.

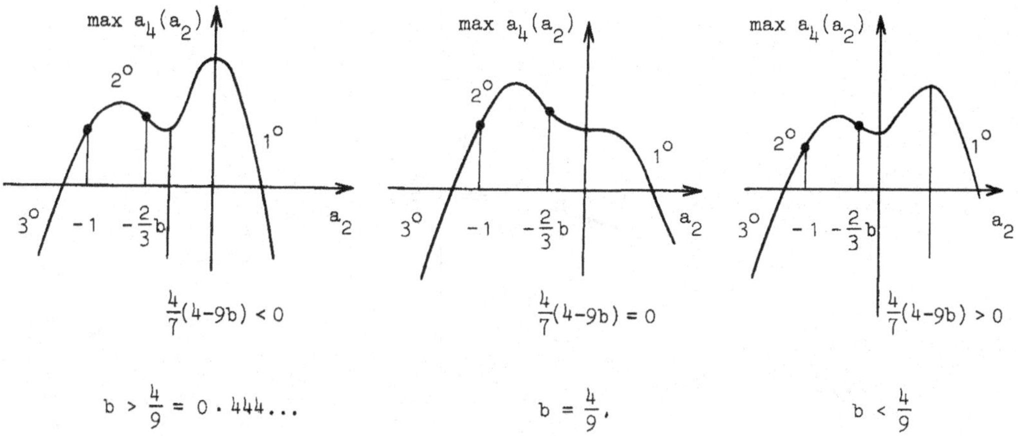

Figure 45.

Figure 45 illustrates $\max a_4(a_2)$ schematically. In Figures 46 and 47 are presented, computer-drawn, the coefficient body (a_2,a_3) with the critical curves for

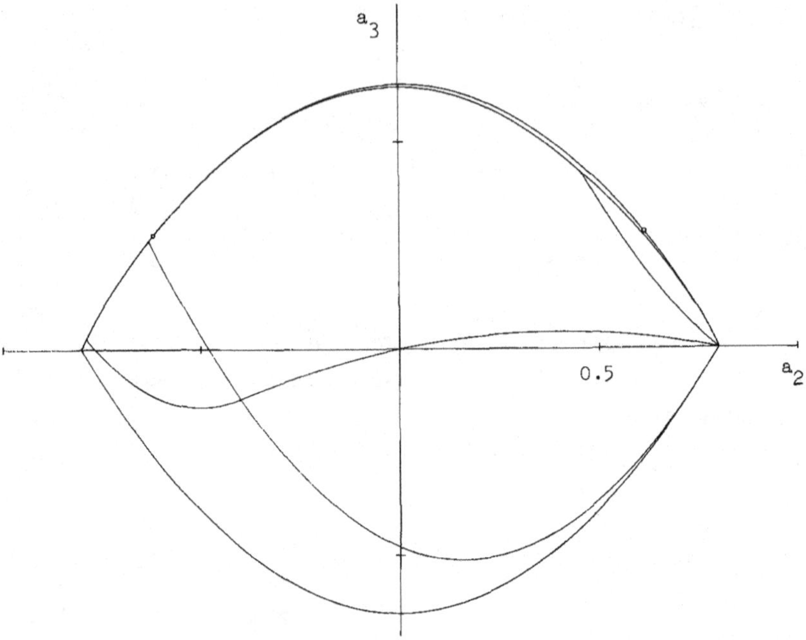

b = 0.6

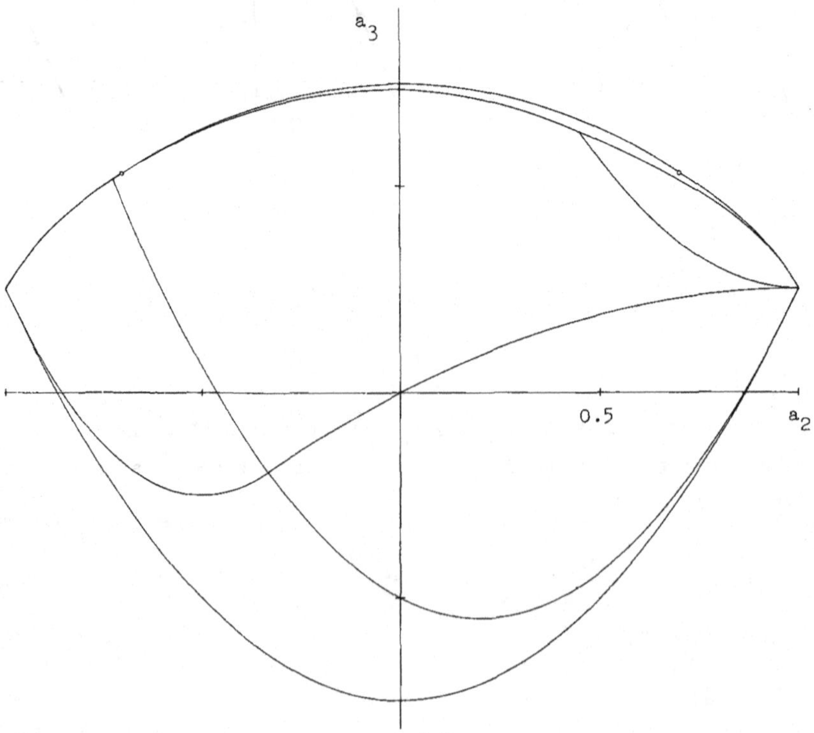

b = 0.5

Figure 46.

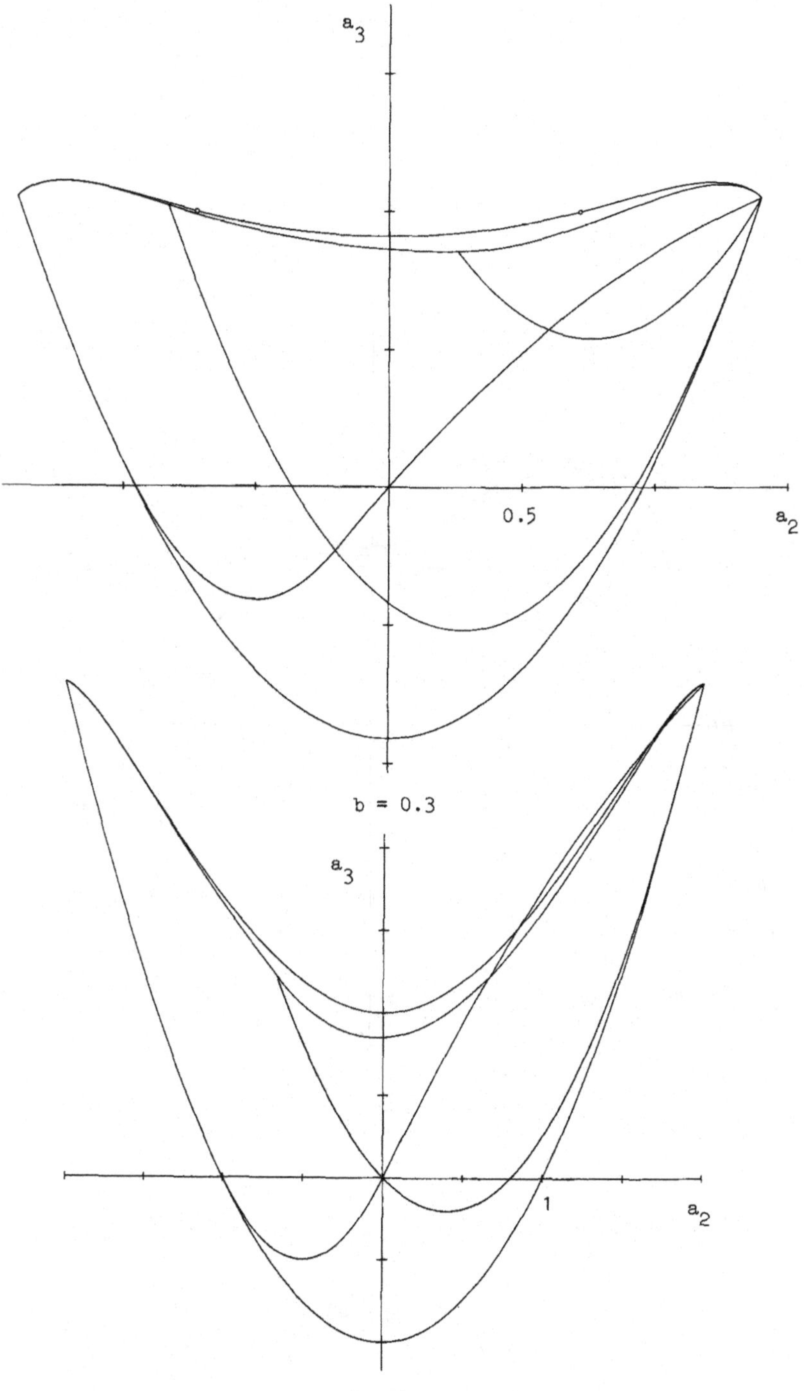

Figure 47.

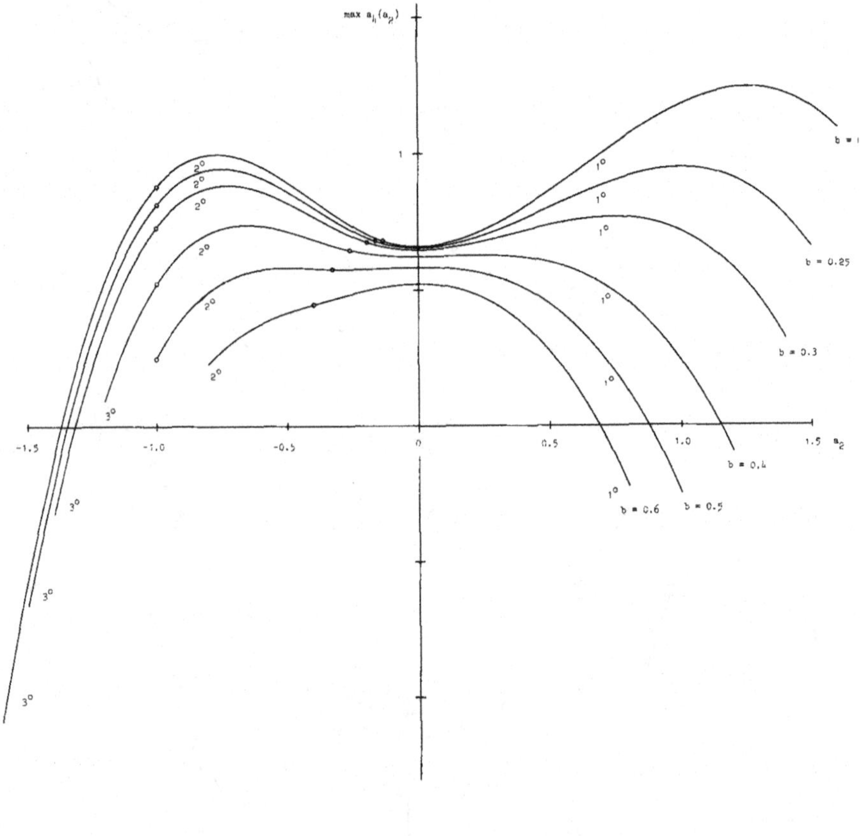

Figure 48.

$$b = 0.6,\ 0.5,\ 0.3,\ 0.$$

In Figure 48 the graphs of $\max a_4\ (a_2)$ are similarly drawn for

$$b = 0.2,\ 0.25,\ 0.3,\ 0.4,\ 0.5,\ 0.6.$$

Observe, that 2° escapes the region II for

$$0.5 < b < 1$$

and 1° similarly escapes the region II if b is small enough. It even lies partly outside the coefficient body (a_2, a_3) for

$$0 < b < 0.092\text{'}548.$$

3. Maximizing a_4 in b.

In III.2.5 of [1] the coefficient a_4 was maximized in $S_R(b)$ for $0.527 \leqq b \leqq 1$. The above Theorem 2/VI.4.2 allows now completing the maximizing of a_4 in b. This is done by comparing the values (13).

Theorem 3. In $S_R(b)$ the coefficient a_4 is maximized in b by aid of the numbers (13):

$$(14)\quad \max_{b\in[1/2,1]} a_4(b) = a_4(b,0;1^\circ) = \frac{2}{3}(1 - b^3);$$

$$(15)\quad \max_{b\in[1/4,1/2]} a_4(b) = a_4\left(b,-\frac{2+\sqrt{4-15b^2}}{5};2^\circ\right)$$
$$= \frac{2}{3} - \frac{2}{5}\left(2+\sqrt{4-15b^2}\right)b^2 + \frac{2}{75}\left(2+\sqrt{4-15b^2}\right)^2;$$

$$(16)\quad \max_{b\in[1/11,1/4]} a_4(b) = a_4\left(b,\frac{4}{7}(4-9b);1^\circ\right) = \frac{2}{3}(1-b^3) + \frac{8}{147}(4-9b)^3;$$

$$(17)\quad \max_{b\in(0,1/11]} a_4(b) = a_4(b,2(1-b);1^\circ) = 4 - 20b + 30b^2 - 14b^3.$$

4. Integration of Löwner's Differential Equation

The generating function ϑ of (6)/VI.3.2 determines the extremal function f through Löwner's differential equation. The integration on the interval $\sigma_2 \leqq u \leqq 1$ is actually performed in VI.1.3 and the result in Theorem 2/VI.1.3 can be directly applied in finding $f_{\sigma_2} = f(z,\sigma_2)$. According to (7)/VI.3.2 the numbers σ_2 and λ are connected so that

$$(18) \qquad b \leqq \sigma_2 = \left(\frac{1-3\lambda}{4}\right)^{2/3} \leqq 1.$$

Thus, it is advisable to express the coefficients $3\lambda - 1 + \sigma_2^{3/2}$ and 3λ, which occur in the equation of f_{σ_2}, by aid of σ_2:

$$3\lambda - 1 + \sigma_2^{3/2} = -3\sigma_2^{3/2};$$

$$3\lambda = 1 - 4\sigma_2^{3/2}.$$

From f_{σ_2} we obtain f_{σ_1} by aid of the right radial-slit-mapping. From this the final $f = f(z,b)$ is obtained by using the left radial-slit-mapping. The result thus assumes the form

Theorem 4. The extremal function $f = f(z,b)$ determined by (6)/VI.3.2 is defined by the conditions

$$(19) \quad \sigma_2^{3/2}(f_{\sigma_2}^{3/2} - f_{\sigma_2}^{-3/2}) - 3\sigma_2^{3/2}(f_{\sigma_2}^{1/2} - f_{\sigma_2}^{-1/2})$$

$$= z^{3/2} - z^{-3/2} + (1 - 4\sigma_2^{3/2})(z^{1/2} - z^{-1/2}),$$

$$(20) \quad \sigma_1^{1/2}(f_{\sigma_1}^{1/2} + f_{\sigma_1}^{-1/2}) = \sigma_2^{1/2}(f_{\sigma_2}^{1/2} + f_{\sigma_2}^{-1/2}),$$

$$(21) \quad b^{1/2}(f^{1/2} - f^{-1/2}) = \sigma_1^{1/2}(f_{\sigma_1}^{1/2} - f_{\sigma_1}^{-1/2}).$$

The two parameters σ_1 and σ_2 lie on $[b,1]$ so that

$$(22) \quad b \leqq \sigma_1 \leqq \sigma_2 \leqq 1.$$

The connection between the parameters (σ_1,σ_2) and the point (a_2,a_3) in the sharpness region II of the coefficient body (a_2,a_3) is obtained from (9)/VI.3.2.

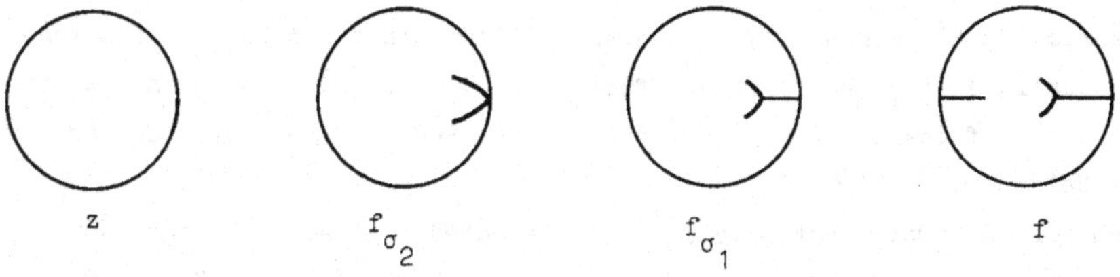

Figure 49.

References

[1] Tammi, O.: Extremum problems for bounded univalent functions. - Lecture Notes in Mathematics 646, Springer-Verlag, Berlin-Heidelberg-New York, 1978.

[2] Haario, H. - Jokinen, O.: On the use of Löwner's functions for finding inequalities for the first coefficient body of bounded univalent functions. - Report of the Department of Mathematics, Ser. A No 9, 1977, University of Helsinki.

[3] Schober, G.: Univalent functions. - Selected topics. - Lecture Notes in Mathematics 478, Springer-Verlag, Berlin-Heidelberg-New York, 1975.

[4] Tammi, O.: On generalizing the Power Inequality for the first coefficient region in the class of bounded univalent functions. - Proc. of the Rolf Nevanlinna Symposium on Complex Analysis Silivri 1976, pp. 103-129.

[5] Kortram, R. - Tammi, O.: Non-homogenous combinations of coefficients of univalent functions. - Ann. Acad. Sci. Fenn. Ser. A I Math., Vol. 5, 1980, 131-144.

[6] Jokinen, O.: On the use of Löwner-identities for bounded univalent functions. - To appear in Ann. Acad. Sci. Fenn.

Index
Part I
(LN Vol. 646)

Index

Part II